The Logic of Environmentalism

First published in 2005 by
Berghahn Books

www.berghahnbooks.com

Library of Congress Cataloging-in-Publication Data

Argyrou, Vassos.
 The logic of environmentalism : anthropology, ecology, and postcoloniality /
Vassos Argyrou.
  p. cm. -- (Studies in environmental anthropology and ethnobiology ; v. 1)
Includes bibliographical reference and index.
ISBN 1-84545-032-9 (alk. paper) – ISBN 1-84545-105-8 (pbk : alk. paper)
 1. Human ecology--Philosophy. 2. Environmantalism--Philosophy. 3.
Ethnobiology--Philosophy. I. Title. II. Series.

GF21.A74 2005
304.2--dc22

British Library Cataloguing in Publication Data

A catalogue record for this book is available from the British Library

Printed in Canada on acid-free paper

ISBN 1-84545-032-9 (hardback)
ISBN 1-84545-105-8 (paperback)

The Logic of Environmentalism
Anthropology, Ecology and Postcoloniality

Vassos Argyrou

Berghahn Books
New York • Oxford

Studies in Environmental Anthropology and Ethnobiology

General Editor: **Roy Ellen**, FBA
Professor of Anthropology, University of Kent at Canterbury

Interest in environmental anthropology has grown steadily in recent years, reflecting national and international concern about the environment and developing research priorities. This major new international series, which continues a series first published by Harwood and Routledge, is a vehicle for publishing up-to-date monographs and edited works on particular issues, themes, places or peoples which focus on the interrelationship between society, culture and environment. Relevant areas include human ecology, the perception and representation of the environment, ethno-ecological knowledge, the human dimension of biodiversity conservation and the ethnography of environmental problems. While the underlying ethos of the series will be anthropological, the approach is interdisciplinary.

Volume 1
The Logic of Environmentalism
Anthropology, Ecology and Postcoloniality
Vassos Argyrou

Contents

Prelude

For a long time, by all accounts the last few centuries, nature was perceived as an intractable domain of utility and danger which, as the language of the nineteenth century would have it, was to be mastered, tamed, brought under 'man's' control, bent to his will, forced to reveal her secrets, compelled to satisfy his needs and minister to his happiness. Such was the 'physics' of the modernist paradigm, the dominant view of the physical world encountered among important people engaged in serious business – people doing science, theorising about the nature and meaning of the world, inventing or using technology, running nations, companies, nations through companies and, of course, empires. The corollary to this vision was that 'man' could, and should do all of the above. For it was only to the extent that he asserted himself in this way that he would fulfil his destiny and become what he was meant to be – the Subject of the world and in control of his destiny. In this vision too, those 'men' who had a different view of themselves and their physical surroundings were perceived, treated and quickly learned to treat themselves as 'primitive' or, as in the postwar, postcolonial lexicon, 'traditional' and 'underdeveloped'. Such was the 'anthropology' of the modernist paradigm, the dominant definition of what it means to be a human being and its humanism – both the belief in the unlimited powers of 'man' and its vision of human unity – which generated the conviction that certain 'men' were entitled, indeed, burdened with the responsibility of mastering Other 'men' in order to humanise them.

In less than three decades, the modernist 'physics' and 'anthropology' have been transformed fundamentally, indeed, in many ways practically reversed. In the paradigm that has now become dominant, nature is neither refractory nor a state – 'the state of nature' – and a predicament. On the contrary, as most people would now say, it is a system of immense complexity that hinges on a precarious balance currently under severe strain, a fragile domain of life that must be urgently protected and cared for, both for its own sake and ours. 'Man' too is no longer the Subject of the world and the indisputable master of nature, but a cautious, sensible and responsible steward. He has been drastically reduced in size and now emerges as the 'human being', a being among other beings in the world and dependent on nature for his every need and very survival. As for those 'men' who were once thought to be 'savage' and 'backward' and in need of enlightenment, they too emerge in a different light. They are now seen as victims of a monumental historical misunderstanding, are portrayed, and have learnt to portray themselves, as living embodiments of an urgently needed ethic of respect for nature, as repositories of a simple yet profound wisdom that the West has long lost in its heedless march for progress. They have been transformed into those who will 'enlighten' the world with this forgotten wisdom and can therefore be called, without the risk of misunderstanding, 'indigenous and traditional peoples'. Such is the environmentalist vision of the world, its 'physics' and its 'anthropology', a vision held with as much conviction and certainty as the modernist vision ever was.

What is one to make of an event of such magnitude and complexity? How is one to understand the emergence of the environmentalist paradigm, its ascendancy and apparent success in such a short time? From which perspective, angle of vision, theoretical orientation should one approach it? There are, to be sure, answers to these questions. One could say, for example, as most people would probably say, that there was little choice in the matter. The threat of an imminent ecological collapse meant that humanity would either mend its ways urgently and drastically or eventually perish. Belatedly, if not too late, practical reason and common sense won the day.

It is true, of course, that there are ecological facts which document the 'environmental crisis' in detail and with precision and although there are also counter-facts – increasingly so – or different interpretations of the same ecological facts, my aim in this book is not to dispute them. To do so, to take sides in the debate over facts would be to reduce environmentalism to a question of scientific objectivity. It would then be possible, hypothetically at least, to prove or disprove ecological facts and, depending on the outcome, to confirm environmentalism in its truth or debunk it as fiction. This would be a rational procedure. Indeed, to a large extent, it is on this assumption that the debate over facts is based – the assumption, that is, of rational actors making informed and rational decisions about the state of the world. Yet things are far murkier and convoluted. Environmentalists and their modernist critics cannot agree on the facts – what is to count as a fact, how those that do count are to be interpreted or what they should mean. And the facts themselves, notoriously, do not speak for themselves and cannot act as final arbiters and guarantors of their truth. Nature, things themselves have long been muted. And although this is only a historical phenomenon that may come to pass, we, in the meantime, have no choice but to speak for them, to listen attentively to their silence and to translate it, arbitrarily but necessarily, into what we think it means. I am well aware that positivists, naïve realists and no doubt environmentalists would disagree with this line of reasoning. There will be, however, many opportunities in this book to revisit the issue and, as I will try to show, although they may not agree, nor can they do without this sort of reasoning.

What I propose to do in this book, then, is to treat the environment as an issue which is neither of the order of truth nor of the order of ideology and false consciousness. I propose to treat it as part of that difficult, dense and ambiguous middle ground that constitutes the realm of culture. My concern is with a prior and more fundamental question than the debate over facts, namely, the question of the facts' own cultural conditions of possibility. Having emerged in specific social contexts, facts circulate and become the object of belief as much as of disbelief, discussion and debate, truths to be upheld or fictions to be rejected. No doubt they are often also a matter of indifference and apathy. The prior question has do with how they emerge and circulate, how their silence is overcome and we get to hear them speak through the inevitable interpreters. What sort of conditions must be fulfilled for ecological facts to emerge and circulate? To be more precise, what sort of cultural assumptions must be in place for these facts to become visible, which is not to say merely observable, identifiable and quantifiable – the greenhouse effect, for instance – but more importantly, value-able, charged with the kind of relevance, significance, gravity and urgency that distinguish the attitude of the activist and the converted from that of the passive observer, sceptic or critic? More broadly, what is the cultural context that renders the environmentalist vision of the world a viable proposition, an issue to be taken up in earnest and pursued in its practical implications and consequences? I do not pretend to have answers to all of these questions or to answer them in detail. But I do attempt an answer, however rudimentary it may be. Nor, it should readily be apparent, am I proposing a causal explanation. My aim is to sketch the conditions and conditionings without which the environmentalist vision of the world could not have emerged and would not have become a serious and legitimate proposition. That it has emerged and has been taken up in earnest cannot be reduced to these conditions partly because of the sheer complexity of the phenomenon and no doubt also because, having emerged, the meaning of environmentalism can neither be dictated nor controlled. As will become apparent in the subsequent discussion on how non-Western governments and bureaucracies understand the 'environmental crisis', it has acquired a life of its own.

The book is based on three guiding assumptions. First, that environmentalism produces a more objectified, totalising and unifying vision of the world than the vision of the modernist paradigm. Second, that to understand the meaning of the former, one has to begin with the meaning of the latter. And third, that what above all is at stake in the environmentalist effort to save nature is not modernist culture, as its apologists fear, but power, the ability of a group of societies to define the meaning of the world for *everyone, yet again*. The book's central theme concerns the apparent reversal of the modernist 'physics' and 'anthropology' in environmentalism, the transformation, that is, of intractable nature into a fragile domain of life and of 'man' the master of nature into the 'human being'. I treat this reversal as a 'phenomenon', a reality experienced and acted upon and hence as something whose truth cannot be questioned or doubted. What is to be doubted rather is the assumption – of environmentalists themselves, their modernist critics as well as the dictates of common sense – that because environmentalism reverses the modernist vision of nature and humanity, it also constitutes a radical rupture with the modernist paradigm. The argument developed in this book is that at a more fundamental level than the phenomenological, environmentalism reflects a return of the same, the reproduction of the same sort of global power relations and the same sort of logic that mark the modernist paradigm at its core.

The Same is a cultural logic and a cultural product. It is a systematic way of imagining and doing, what above anything else defines the modernist subjectivity and makes it what it is – a subjectivity that constantly strives for unity, purity and innocence because it can find its rationale and reason of existence nowhere else except in such a vision of the world. The Same of the modernist paradigm is Humanity and its humanism, which is epitomised in all those attempts to efface, by proving groundless, the divisions of the Same – divisions based on class, gender, race, ethnicity and cultural difference, to mention only the major ones. The Same of environmentalism, in turn, is the Same of the modernist paradigm incorporated into, but in no way negated by, a grander domain of unity, purity and innocence. And it is reflected in the environmentalist effort to efface the greatest of all modernist divides – the division between Humanity and Nature – as in the claim that 'all is One', for instance, or, in the less radical, more 'mainstream' environmentalist version, that 'we are part of nature and nature is part of us'. Environmentalism reflects a return of the Same because it operates on the basis of the modernist logic. In this non-phenomenological and more profound sense, it differs from the modernist paradigm only to the extent that it takes the logic it has inherited to its logical and onto-logical conclusion. I should perhaps reiterate that this argument is not meant as a causal explanation of environmentalism. No doubt the environmentalist vision of the world could not have emerged without the underlying assumptions that constitute the logic of the Same. Yet this is not to say that it can be reduced to it. It would be more accurate to say that environmentalism was an immanence, a cultural possibility inherent in the modernist paradigm triggered by the conjunction of specific historical circumstances and events, a virtual reality waiting for actualisation.

Environmentalism reflects a return of the same of the modernist paradigm in another, equally important sense. As I have already suggested, this is the same of power, the ability of a group of societies to define and redefine, construct and reconstruct the order of the world and the world order. Environmentalism repeats the historical gesture that marked the colonial enterprise and its civilising mission. The rest of the world is once again presented with a new reality – presented, that is, *fait accompli* – and is expected, cajoled, encouraged, assisted, threatened to take a stance and come to recognise it as such a reality. And just as surely, the rest of the world responds in the manner that marked its confrontation with and accommodation of the modernist vision during colonial times. It acts suspiciously – for it is still burdened with the legacy of modernism – doubts, questions, rejects, negotiates, moves strategically and tries to gain advantage, co-opts, recognises, endorses. It engages, in short, in that wide range of practices which, to paraphrase Bourdieu, constitute the complicity that unites the rest of the world, as the 'Rest', with the West in disagreement or, what is another way of saying the same thing, the disagreement that divides it and the West in complicity. The 'Rest' does everything that needs to be done – unwittingly and unwillingly no doubt – to ensure that it remains locked in that relationship which ties it to the West and defines it as an irredeemable Otherness. Which is not to assign blame or to suggest that anyone has escaped this predicament and can talk about it from the outside.

This book, then, is about how the world changes but remains the same – despite phenomena to the contrary, and all good intentions notwithstanding. Not a very profound thought perhaps, but at the threshold of what promises to be another two hundred years of (post)colonial solitude, surely a necessary one.

Vassos Argyrou

1 First Change

'The Idea of Nature'

'In the history of European thought', wrote R.G. Collingwood in his history of *The Idea of Nature*, there have been 'three periods of constructive cosmological thinking'. The first was the age of classical Greece, the second the sixteenth and seventeenth centuries and the third the nineteenth century. During these periods, Collingwood goes on to say, 'the idea of nature [came] into the focus of thought, [became] the subject of intense and protracted reflection' and eventually produced a new vision of reality (1945: 1).

Collingwood was writing in the 1940s and could not have predicted that European thought would be preoccupied with the idea of nature again only three decades after the publication of his book. He was writing at a time when concern about the state of nature was practically non-existent and cannot be reproached for ignoring the impact of the idea of nature on nature. We, by contrast, are confronted with the latest preoccupation of European thought and are constantly reminded of the state of (mainly non-European) nature. We can ignore neither. But there is another difference between Collingwood's time and ours. He was writing during the end of an era, the time when Empire was coming to a close and therefore far too early for anyone to even consider the long-term impact of European thought on the emerging postcolonial societies. We, by contrast, are writing decades after the collapse of Empire and, having lived through this period of postcoloniality, we have come to realise that in a certain fundamental sense it has never left us, that it was as much an Empire of the mind as of anything else, that the webs of meaning and power that European thought has spun over the last two centuries still hold the rest of the world captive. If, then, we cannot afford to ignore the latest preoccupation of European thought, neither can we afford to take it for granted. We must question and problematise it, even if, in a paradoxical sort of way, this attitude too confirms the firm grasp of European thought over everyone's imagination.

The Greek view of nature, Collingwood goes on to say in his history, was based on an analogy – the earth as an organism. Ancient Greeks regarded nature as a universe of bodies in motion; motion itself was due to vitality, which made nature an entity with a soul, a living and feeling organism. Moreover, despite incessant movement, there was regularity and order in nature, both of which were regarded by ancient Greeks as the unmistakable sign of mind or intelligence. Hence, nature 'was not only a vast animal with a "soul" or life of its own'; it was also 'a rational animal with a "mind" of its own' (1945: 3). It was a living, feeling and thinking organism that strove to achieve its own end or *telos*. The ancient Greek view was displaced by what Collingwood calls the 'Renaissance view of nature'. This view was also based on an analogy, but one that drained all life and intelligence from the animistic world of the ancient Greeks. For the European philosophers and scientists of the sixteenth and seventeenth centuries, nature was like a machine, a view made possible both by the belief in a 'creative and omnipotent God [and] the experience of designing and constructing machines' (1945: 8). From here, Collingwood points out, it was 'an easy step to the proposition: as a clockmaker or millwright is to a clock or mill, so is God to Nature' (1945: 9) – that nature was God's artefact in much the same way as a clock is a clockmaker's creation.

The last period, Collingwood calls 'modern'. During this time, the nineteenth century, a new idea of nature developed that drew inspiration from an unlikely source – history. The analogy in this case was between the 'processes of the natural world as studied by natural scientists and the vicissitudes of human affairs as studied by historians' (1945: 9). By this time, evolutionism had become the dominant paradigm and human affairs were perceived in terms of progress and development – an idea 'derived from the principle that history never repeats itself'. In a similar vein, nature became a domain in which 'nothing is repeated, a second world of progress characterised … by the constant emergence of new things' (1945: 14). Nature as an evolving and progressing entity could not be a machine, however. But in the disenchanted universe of the nineteenth century, nor could it be what it was for the ancient Greeks, an entity with a soul and mind of its own. Nature was from now on perceived the way in which Darwin himself perceived it: as a terrain 'of an endless succession of experiments … to produce organisms more and more intensely and effectively alive', but an entity nonetheless 'wholly devoid of conscious purpose' (1945: 134–35).

Such, in a nutshell, is Collingwood's history of the idea of nature in European thought up to the middle of the twentieth century. Let us note a few things about this intellectual history. First, the ironies. Although the age of classical Greece was a period of 'constructive cosmological thinking', in the end the Greek idea of nature emerges in Collingwood's book as a construct of spurious value: 'That vegetables and animals are physically akin to the earth is a belief shared by ourselves with Greeks; but the notion of a psychical and intellectual kinship is strange to us, and constitutes a difficulty in the way of our understanding the relics of Greek natural science which we find in the literature' (1945: 4). What is one to make of this inconsistency? If we can hardly understand the 'relics' of Greek natural science, if Greek assumptions about the nature of nature are 'strange' to us, why does Collingwood count this as a period of 'constructive cosmological thinking'? In what sense is it 'constructive' – what does it construct? – and in relation to what? Collingwood provides no answers. Second irony: thirty years after the publication of Collingwood's book, Greek cosmological thinking becomes constructive once again but for rather different reasons from those that Collingwood may have had in mind. To begin with, the Greek view of nature is no longer strange to us. We encounter it often in the course of everyday life and have become quite familiar with it. Moreover, and more importantly, for many people it constitutes a core assumption about the nature of nature and is said to provide the only hope for the future of the planet. As we shall see, the Greek earth goddess, *Gaia* (Lovelock 2000), the living organism and thinking being, has been resurrected and reigns over 'barbarians' and Greeks alike.[1] Third irony: the 'relics' of natural science are no longer Greek but western European. They are to be located in particular in the literature of the sixteenth and seventeenth centuries that we have no difficulty in understanding and even less in condemning. The 'constructive' cosmological thinking of this period, which produced the idea of nature as a mechanical entity, has been denounced by everyone concerned with nature for being thoroughly destructive.

The irony in this state of affairs is not that ideas change, even fundamental ideas such as the idea of nature. Nor is it, as in this case, that they are often completely reversed. Nor is it again that they can be reversed in a very short period – in this case, three decades. The irony, rather, has to do with the economy of memory, the convenient disregard of history and culture in understanding, the naïveté of activist conviction – which, as I shall try to show in this study, is founded in compelling ontological reasons – and perhaps above all, the fanciful expectation that the latest incarnation of the idea of nature, the environmentalist idea itself, can pass as the only true picture of reality – which, judging by its success, may not be such a fanciful expectation after all.

The second thing to note about Collingwood's history is the three-stage evolutionary schema of European progress that it faithfully reproduces – classical Greece, Renaissance, Modernity. Although the ideology that underpins this schema has been exposed many times, I shall nonetheless deal with it here, even if briefly, in order to link both with the questions raised above about Collingwood's inconsistency and the claims to be made below. The first thing to draw attention to is the obvious – the yawning gap between the age of classical Greece and the Renaissance. Since this is a history of progress, the history of Christianity and of the 'Dark Ages' in particular has no place in it. Let us also note that if classical Greece is removed from the equation, the history of

European progress and therefore the age of 'the Europe' that Europe wishes to project for itself to its Others and itself is reduced to a mere four centuries. This is not to say only that Europe suddenly emerges with rather shallow roots. It is also to say that it emerges with roots that may have been drawing sustenance from Other soils outside Europe. As Bernal (1987) first suggested, with the hardening of European racism in the eighteenth century, this possibility became unacceptable. The invention of classical Greece as the cradle of European civilisation did much to sustain the fiction of European purity, if not to the present day, certainly well into the second half of the twentieth century.[2] The need for cultural purity, or at any rate, the need for cultural distinction goes a long way in explaining Collingwood's untenable position – the claim that the age of classical Greece was a period of 'constructive' cosmological thinking and the simultaneous denunciation of Greek natural science as a 'relic' of the past that we, today, can hardly understand. What the Greek idea of nature signifies or, to be more precise, what its inclusion in 'the history of European thought' indicates, is not so much a period of constructive cosmological thinking as a period of disjunctive ideological thinking during which Europe sought to define itself in opposition to, and hence to distance and distinguish itself from, its Others. The construct which classical Greece helped to produce was not so much the idea of nature (or the idea of democracy, art, philosophy, rationality, science and so on) as the idea of Europe itself – European identity.

The idea of nature that was to become dominant during the last few centuries – let us call it 'modernist' – is nonetheless implicated in the idea of Europe far more than any other idea. As environmentalists are never tired of pointing out, it was not simply an idea but also, inextricably, a specific attitude towards the physical world. Whether a machine, as in the sixteenth and seventeenth centuries, or a domain of life mechanistically producing new versions of itself, as in the nineteenth, nature came to be perceived also as an intractable domain of danger and, above all, utility to be mastered by 'man' and brought under his control. As Heidegger (1977a) complains in his critique of science and technology, nature was to be reduced to a 'standing reserve', a stockpile of resources for the taking of 'man'. This ontological reduction of nature was to a large extent responsible for both the physical and ideological expansion of Europe. It no doubt facilitated European overseas expansion. But it also consolidated European power and made possible the colonisation of the rest of the world. Moreover, and to the extent that mastery of nature came to be seen as the most remarkable achievement of European 'man', it provided the most tangible proof of European cultural superiority and solid ideological grounds for the legitimisation of colonialism. Mastery of nature came to be seen as the unmistakable mark of civilisation, the core characteristic not of European 'man' but 'man' as such. To paraphrase Marx who expressed this idea better than any bourgeois thinker, 'man' makes himself and is himself only insofar as he remakes the world around him. The more he changes the world around him, the more he becomes his true self. Such was the modernist definition of humanity – what it means to be a human being, a society, a nation, a society of nations – the 'anthropology' that lent support to the modernist 'physics' of nature and was in turn supported by it, and the vision of the world deeply implicated in the politics of colonialism. No wonder then that those who justified colonialism 'were firmly convinced that they were acting in the long-term interests of the peoples brought under European rule' (Adas 1989: 200). According to this logic, those people that had not yet mastered nature did not because they could not – rather than because they may have had a different vision of the world – because they were innocent, ignorant and superstitious. It befell European 'man' who knew better to rectify these shortcomings. That was his 'civilising mission', the burden he was destined to bear in the colonies as a government functionary, a military man, trader, ideologue and, of course, missionary.

Beyond the history of the idea of nature and the history of its mastery by European 'man', then, there is another story to be told. This is the story of the mastery of the rest of the world by a handful of European nations in which both the idea of nature and the mastery of the entity to which the idea refers are deeply implicated. These stories must be told for the same reason that all stories are usually told – because we should not forget to remember. We should not forget to remember, to begin with, that the modernist perception of nature and humanity was held to be a true representation of reality with as much conviction and certainty as environmentalists hold their own views today. Unless one is prepared to argue, as environmentalists often do, that the modernist perception was little more than a fateful 'misconception' of reality – in which case one must explain the persistence of such a 'misconception' for almost three centuries in a culture dedicated to eradicating all fictions and illusions – unless one is prepared to do so, recalling the modernist certainties may bring once again into the 'focus of thought' the role that history and culture play in shaping any perception of the world, including, no doubt, the environmentalist. This sort of remembering would not merely throw into relief the naïveté of both the modernist and environmentalist convictions. It would also open up a new domain of inquiry in which this naïveté could become intelligible.

For what it is worth, we should also not forget to remember the historical responsibilities that burden the present – the fact that the modernist 'physics' and 'anthropology' set the rest of the world onto a historical trajectory that now emerges as little more than a fateful mistake, a road that leads nowhere anyone would wish to go. In light of this revelation, we should remember the stigma attached to those outside this history, their forceful incorporation and, as we shall see below, until very recently, the pressures exerted on them to speed up the pace of reform and to 'leap across the centuries' that separated them from Europe's present. We should remember all this every time environmentalists call on Others to abandon the modernist trajectory that leads nowhere and to joint their train of thought and practice that allegedly goes everywhere. If nothing else, it could help keep things in some kind of perspective. But by the same token, we should not forget either the historical complicities that burden the present, the logical conformities and paradoxical empowerments that have led Europe's Others to recognise the stigma attached to them – which is not to say necessarily accept it – to will their incorporation into the world history and in all but very few cases to commit themselves wholeheartedly to the proposed temporal 'leap'. It may sensitise us to its uncanny resemblance to the current historical conjuncture, the new round of logical conformities, empowerments, recognitions, complicities and, no doubt, entrapments.

Finally, we should bear in mind that unlike the otherness of nature, no one quite knows what needs to be done about the Otherness that plagues Europe's Others. No doubt many have suggested remedies, both within the West and outside it: from anthropologists, to liberal politicians and intellectuals, to contemporary poststructuralists whose aim is nothing less than to 'de-centre' the West itself; from the early anti-colonial nationalists such as Gandhi, to subsequent activists and theorists like Fanon, to present-day 'postcolonial' scholars. Yet no one has ever succeeded in extricating them from the webs of power that Europe has spun over the centuries. What is more, no one quite knows whether anything of the sort is possible at all. If the history of anthropology and other liberal discourses is anything to go by, it seems that every attempt to liberate them has done little more than to entangle them further and trap them even more securely.[3]

For all these reasons, we – Europeans, non-Europeans and whatever 'hybrid' personas may be 'between and betwixt' – should not forget to remember. Let us, then, proceed with these stories and the ways in which they intertwine and overlap over the centuries. Let us go back in time and try to trace the development of these events, if not in detail – for that would require a different, properly historical study – at least by highlighting the key ideas and practices, underlining the important shifts, drawing attention to fateful moments and pronouncements, referring to major names and major namings. We must revisit the period when nature and humanity were not in the 'focus of European thought' and the subject of 'intense and protracted reflection', constituted different sort of entities, subsisted at the periphery of relevance and concern. This would be the time also when different categories were involved in identification and classification, differentiation and distinction, when the meaning of the world itself was different. We must then try to trace the change itself: the transformation of nature into an object to be mastered and brought under 'man's' control; the transformation of 'man' into the master of nature and the positing of this idea as the essence of what it means to be 'man'; the related transformation of the means of transforming nature – machines – into 'the measure of men' (Adas 1989);[4] and finally, the transformation of non-European 'man' into a semblance of the sort of 'man' that European 'man' had in mind. Having highlighted the contours of this change in the order of the world, we must turn to the second change – the change that is currently upon us, is pulling the entire world along with it and looks destined to lead us into another long period of 'progress' and (post)colonial solitude.

'If we have no rivers, we make canals'

It would hardly be possible to pinpoint historically the emergence of this series of transformations. There is no such point in time, except from the perspective of a different time, the perspective, that is, of a detached observer who can totalise and fix events with the benefit of, among other things, hindsight.[5] Let us, then, begin with the early centuries of European overseas expansion and proceed with as broad and chronologically non-specific signposting as possible. As Adas (1989: 6–7) points out, during this early period,

> European judgements about the level of development attained by non-Western peoples were grounded in the presupposition that there are transcendent truths and an underlying physical reality which exists independent of humans, and that both are equally valid for all peoples. ... The Europeans' sense of superiority was anchored in the conviction that because they were Christians they best understood the transcendent truths. ... Right thinking on religious questions took precedence over mastery of the mundane world in setting the standards by which human cultures were viewed and compared.

The European sense of superiority was not entirely based either on religion or on the mastery of the mundane world. As Pagden (1982) shows in his study of the European encounter with American Indians, the learned of the time were thoroughly steeped in the Aristotelian view of the world which defined civility on the basis of several other characteristics: life in the city, which is to say, a socially ordered and organised life, form of marriage and family, commerce, customs and so on. Above all, and the prerequisite for these characteristics, was mastery of internal nature, control of human passions and desires through the use of reason. Characteristics of civility not directly related to religion become apparent also in the negative instance, in terms, that is, of what struck the educated European élite as being absent from non-European societies. On the basis of accounts from Montaigne, Boemus and Le Roy, Margaret Hodgen (1964: 199) produced the following list:

> No ... letters; no ... laws; no ... kings or magistrate, government, commonwealth, rule, commanders; no arts (or occupation); no traffic (or shipping, navigation); no husbandry (or agriculture, tillage, tilth, vineyards, sowing or planting); no money (or no exchange, money, riches); no cloths ... no marrying (no wedding, no respect of kindred).

In their judgements of themselves and Others, then, Europeans employed several criteria. There is little doubt however, that religion was the 'most vital of all the requirements for civility' (Pagden 1982: 77). Writing in 1575 about the discovery of the New World, Le Roy has this to say about what was found to be missing among native populations and what not: 'they which have navigated thither, have found people living yet as the first men, without letters, without Lawes, without Kings, without common wealthes, without arts; but yet *not without religion*'.[6] Natives may have been leading the lives of 'the first men', but even the first men knew of some sort of religious life.

We do not know Le Roy's sources. Many who had 'navigated thither', however, often reported the contrary, namely, that although some Indian tribes, most notably in central and south America, were religious, the vast majority were not. If anything, the reports pointed out, natives were given over to idolatry and led lives of immorality and hedonism (Pagden 1982). Indeed, accusations of lack of religion were to continue well into the nineteenth century, no doubt partly as a result of the hardening of the divisions between Europe and its Others which, by this time, had taken the form of an entrenched racism. Thus, Sir Samuel Baker, the famous British explorer, found that the Nilotic peoples were 'without exception … without a belief in a supreme being … [or] any form of worship or idolatry; nor [was] the darkness of their minds enlightened by even a ray of superstition'.[7] And J.D. Lang, writing about Australian aborigines, declared that they had 'no idea of a supreme divinity, creator, and judge, no object of worship, no idol, temple, or sacrifice … "nothing whatever of the character of religion to distinguish them from the beasts that perish"' (Tylor 1874: 418). Such accusations were widespread and part and parcel of what Stocking (1987: 153) calls the 'apocryphal nineteenth-century ethnographic account: "manners, beastly; religion, none"'. Apocryphal or not, lack of religion was a serious accusation. To believe in the wrong God, like 'infidels', was perhaps understandable; to worship idols, like pagans and 'barbarians', may have been excusable; but to have no religion at all was a different matter altogether, a sign of something fundamentally wrong. As the quotation from Tylor makes clear, it placed in serious doubt one's very humanity. Religion, Le Roy argued in 1575, 'is more natural to man than all his other arts and inventions'.[8] If, as Le Roy and his contemporaries insisted, 'man' is by his nature a religious being, lack of religion suggested beings that may have looked like, but were not quite, men.

The importance of religion as a sign of humanity and hence of some sort of civilised existence is graphically captured in the history of anthropology itself. And so is, in fact, the shift from religion to science as the primary means by which Europeans distinguished themselves from Others. As I have argued elsewhere (Argyrou 2002), even as late as the nineteenth century, sympathetic observers were very much concerned with proving wrong all those Europeans, such as explorers and traders, who accused natives of having no religion at all.

Tylor, for example, the paradigmatic figure of Victorian anthropology, argued that such Europeans could not perceive the religious side of native life because they understood religion in the restricted sense imposed by their own culture. To remedy the situation, Tylor defined religion in the broadest possible terms – as belief in spiritual beings. By the middle of the twentieth century, the critical issue for those who sought to redeem Others from the calumny of inferiority was no longer the existence of religion in native societies but the existence of science. That natives were religious was no longer in doubt partly because religion was no longer a critical element in the definition of cultural value and worth. What counted now above anything else was science and rationality. This is graphically portrayed in, among others, the work of one of the major anthropological figures of the twentieth century, Lévi-Strauss. In the *Savage Mind*, for instance, Lévi-Strauss sets out to demonstrate the existence of science in native societies by employing Tylor's own strategy, by using, that is, broad and elastic definitions. As is well known, for Lévi-Strauss, native science is a science of the concrete and although strategically different from western science, which is a science of the abstract, it is neither historically prior nor in any way inferior to it. The aim of both is to classify the world, and both do so equally effectively in their different ways.

Between Tylor and Lévi-Strauss stands the transitional figure of Malinowski – transitional both in terms of anthropology's change in orientation, shifting as it did from evolutionary to synchronic studies, and in terms of the relative weight placed on religion and science as measures of cultural value and worth. Hence, the opening sentence in Malinowski's famous essays on magic, science and religion in which natives emerge as both religious and scientifically minded: 'There are no peoples however primitive without religion and magic. Nor are there, it must be added at once, any savage races lacking either in scientific attitude or in science, though this lack had been frequently attributed to them' (1954 [1925]: 17). Read out of context, this would be a rather puzzling statement, not least because we take it for granted that for most people it is science that counts, not religion and magic. Yet Malinowski had scores to settle with observers of native life both of the past and of his own time.

To return to Tylor, by the time he was writing, the second half of the nineteenth century, religion was hardly a significant basis for comparison between Europe and its Others. His aim in locating religion in native societies was primarily to undermine the dehumanising effects of the racist discourse of many of his contemporaries. For Tylor, natives were indeed primitive but their cultural inferiority did not make them less than human. The existence of religion was fundamental proof of the natives' humanity. In tune with the spirit of his age, Tylor would use not religion but science and technological achievement as the critical factors in the evaluation of societies. As Adas points out (1989: 144), although 'most nineteenth-century observers mixed nontechnological or nonscientific gauges – systems of government, ethical codes, treatment of women, religious practices, and so on – with assessments of African and Asian material mastery', this pluralism in evaluative criteria was not to last for very long. 'As the century passed … colonial administrators and missionaries, travellers and social commentators increasingly stressed technological and scientific standards as the most reliable basis for comparisons between societies and civilizations'.

It is important to emphasise that mastery of the mundane world was by no means an uncontested view of the meaning of either nature or humanity in nineteenth-century Europe. As Berlin points out, the Romantics did not conceive of the physical world as a 'set of facts, as a pattern of events, as a collection of lumps in space, three-dimensional entities bound together by certain unbreakable relations, as taught to us by physics, chemistry and other natural sciences'. They understood it instead as a 'process of perpetual forward self-thrusting, perpetual self-creation' – a process essentially beyond rational human knowledge and outside human control. Not that all Romantics agreed on the nature of this 'self-thrusting' of the physical world. For some, like Schopenhauer, it was hostile, which meant that it could not but 'overthrow all human efforts to check it, to organise it, to feel at home in it', despite short-term successes. For others, less pessimistic than Schopenhauer, it was a friendly process. 'By identifying with it … by discovering in yourself those very creative forces which you also discover outside … by seeing the whole thing as a self-organising and self-creative process, you will at last be free' (Berlin 1999: 119–20). For the latter kind of Romantics, then, as much as for many environmentalists, the attempt to control and master nature reflected a fundamental misunderstanding of both the nature of nature and the nature of humanity. Humanity was part of nature, one aspect of this self-thrusting, self-unfolding, self-creative force. Instead of trying to control it, humanity should re-immerse itself in nature and allow itself to be carried by the current. No wonder, then, that, as we shall see, Romantics of this persuasion, such as Herder, were staunch critics of European overseas expansion and the 'civilising mission' of colonial officials and missionaries. They too, much like environmentalists, imagined indigenous populations as the living embodiment of their idea of nature.

But nor should the persistence of Romanticism in the midst of the more utilitarian understanding of the world be surprising. Hegemonies generate counter-hegemonies (Williams 1977), worldviews become orthodox only when heterodox views emerge on the scene (Bourdieu 1977, 1990). If anything, Romanticism testifies to the strength of the utilitarian vision, its dominance over other ways of understanding the world. Outside Romantic circles, humanity and nature were kept meticulously apart, the assumption being, as we have seen, that humanity would become itself, that is, fully human or at any rate civilised, not by immersing itself in nature but, on the contrary, only insofar as it succeeded in extricating itself from it as much as possible. This not only meant turning external nature into an object to be manipulated for human needs. It also meant, inextricably, bringing under control internal nature, the passions, instincts and, as David Hume made clear in the eighteenth century, the easily excitable and therefore unreliable human imagination.[9] In Victorian England, Stocking (1987: 36) points out, civilisation 'tended to be seen as a triumph over rather than an expression of the primal nature of man, just as it was a triumph over external nature'. Mastery of internal nature was regarded as the precondition for mastery of the physical world. As Hume (1977 [1748: 79]) put it a century earlier, 'gross and vulgar passions [leave] little room for reason or reflection'. They inhibit the understanding and give rise to beliefs about 'supernatural and miraculous relations [which] are observed chiefly to abound among ignorant and barbarous nations'. Understanding of *natural* relations, of the way in which the material world operates, on the other hand, required the control of passions and the development of reason.

It seems, then, that there was little room in the dominant ideology for the 'natural man' and the 'noble savage' of Romantics of the likes of Rousseau and Herder. As Hamilton (1969 [1830]: 14–15) put it, criticising 'Rousseau, &c.', the Romantic calls 'to embrace the cause of savage nature ... have never made any progress'. The idea was 'too repugnant to the sentiments which men imbibe in an advanced state of society ... . It may float lightly on the imagination of a few, but it will not alter their character, or influence their conduct'. Nineteenth-century European 'man' perceived himself to be thoroughly cultural, in control of his body and his passions and through the power of his reason and his physical power, increasingly invested in machines, in control of external nature as well. 'Man' was not born free only to end up everywhere in chains, as Rousseau lamented. For the nineteenth-century bourgeoisie, he was born in chains, subject to his instincts and the forces of external nature, only to become free by mastering both himself and the physical world around him. Lubbock, one of the leading Victorian anthropologists, expressed this widespread sentiment in no uncertain terms:

> The true savage is neither free nor noble; he is a slave to his own wants, his own passions; imperfectly protected from the weather, he suffers from the cold by night and the heat of sun by day; ignorant of agriculture, living by the chase, and improvident in success, hunger always stares him in the face, and often drives him to the dreadful alternative of cannibalism or death.[10]

Nor does there seem to have been much room in the dominant European ideology of the nineteenth century for earlier civilisations, such as those of Mesopotamia, India or Egypt. As Henry Thomas Buckle (1878: 50) pointed out in his monumental *History of Civilisation in England*, although the civilisation of

'Asia and Africa … was the earliest, it was very far, indeed, from being the best or the most permanent'. It was neither the best nor the most permanent because unlike the European civilisation of Buckle's time, it was not based on 'man's … own resources', namely, his reason and 'that bold, inquisitive, and scientific spirit, which is constantly advancing, and on which all future progress must depend' (1878: 131). Although the civilisation of Asia and Africa emerged in the most fertile lands, which guaranteed abundant returns, it soon stagnated and eventually withered away precisely because 'the powers of nature, not withstanding their apparent magnitude, are limited and stationary'. What was needed, in addition, to ensure continuous growth were the powers of 'man', which were unlimited. In the humanistic spirit of his age, which is no doubt still with us, Buckle could see no evidence 'which authorises us to assign even an imaginary boundary at which the human intellect will, of necessity, be brought to a stand' (1878: 50–51). The key to progress, then, which the Europe of Buckle's time had discovered, was the combination of the enormous but limited powers of nature and the even greater and unlimited power of 'man'. This is precisely what was missing from Asia and Africa. Apparently, at this early stage of civilisation, 'man' knew nothing of the unlimited capacities of the human intellect. His understanding, Buckle says, recycling the empiricist critique of internal nature, was 'too weak to curb the imagination and restrain its dangerous licence' (1878: 119) – dangerous licence because the imagination encouraged 'superstition and discouraged knowledge'. If we look at 'the history of the world as a whole', therefore, we can see clearly that 'the tendency has been, in Europe, to subordinate nature to man; out of Europe, to subordinate man to nature'. Such is the rule, and although 'there are, in barbarous countries, several exceptions … in civilised countries the rule has been universal' (1878: 151).

European 'man's' ability to subordinate nature explained European superiority over the rest of the world. It is worth quoting from Buckle at some length here:

> If … we take the largest possible view of the history of Europe, and confine ourselves entirely to the primary cause of its superiority over other parts of the world, we must resolve it into the encroachment of the mind of man upon the organic and inorganic forces of nature. To this all other causes are subordinate … .
>
> It is accordingly in Europe alone, that man has really succeeded in taming the energies of nature, bending them to his own will, turning them aside from their ordinary course, and compelling them to minister to his happiness, and subserve the general purposes of human life. All around us are the traces of this glorious and successful struggle. Indeed, it seems as if in Europe there was nothing that man feared to attempt. The invasions of the sea repelled, and whole provinces, as in the case of Holland, rescued from its grasp, mountains cut through and turned into level roads; soils of the most obstinate sterility becoming exuberant, from the mere advances of chemical knowledge; while in regard to electric phenomena, we see the subtlest, most rapid, and most mysterious of all forces, made the medium of

thought, and obeying even the most capricious behest of the human mind. In other instances, where the products of the external world have been refractory, man has succeed in destroying what he could hardly hope to subjugate. The most cruel diseases ... have entirely disappeared from the civilized parts of Europe; and it is scarcely possible that they should ever again be seen there. Wild beasts and birds of prey have been extirpated, and are no longer allowed to infest the haunts of civilized men. [What is more] in our age of the world, if nature is parsimonious, we know how to compensate her deficiencies. If a river is difficult to navigate, or a country difficult to traverse, our engineers can correct the error, and remedy the evil. If we have no rivers, we make canals; if we have no natural harbours, we make artificial ones. (1878: 153–56)

Nothing less than a paean to the unlimited powers of 'man' – his rationality and scientific spirit, knowledge and understanding, boldness and inquisitiveness, his mastery of, and hence freedom from nature, want, disease, poverty. A paean to 'man' and a devastating blow to nature – its obstinate sterility, its refractoriness, cruelty, wildness, infestations and diseases, parsimony, deficiencies, its errors and evils – and no doubt a nightmare vision for environmentalists. A paean to *European* 'man' and a devastating critique of every Other 'man' – his excitable imagination, the feebleness of his mind, his ignorance and superstition, his inability to bend nature to his will and compel it to minister to his own happiness – the Other 'man' whose life was at the mercy of both internal and external nature.

There is, of course, little doubt that Buckle's account is highly rhetorical but not for this reason any less representative of the spirit of his age. Buckle himself may not have been a figure of distinction in the European intellectual tradition, but as Bury (1932: 309–10) points out, his *History of Civilisation in England* 'enjoyed an immediate success [and] did a great deal to popularise' the ideas of such important figures as Comte and J.S. Mill. Indeed, Buckle cites all six volumes of Comte's *Courts de Philosophie Positive* and summarises the essence of his own argument in the form of a quotation from Mill's *Principles of Political Economy*: 'Of the features which characterize this progressive economical movement of civilized nations, that which first excites attention ... is the perpetual, and, so far as human foresight can extend, the unlimited growth of man's power over nature'.[11]

'Man's' subordination of nature, then, emerges as a powerful image in the European imagination of the nineteenth century. Moreover, if the quotations above are anything to go by, it emerges also as the single most important criterion in defining the meaning of humanity, the nature of civilisation, the essence of cultural value and worth. This image was hardly confined to scholars and the educated middle classes. As Buckle himself points out, evidence of the 'glorious' struggle against nature and the fruits of this struggle were everywhere for everyone to see: roads, canals, harbours, bridges, electricity, railways, factories – an unprecedented transformation of the European landscape that Europeans and non-Europeans alike could not help but notice. Bury (1932: 324) concurs: 'The spectacular results in the advance of science and mechanical technique brought home to the mind of the average man the conception of an indefinite increase of man's power over nature as his brain penetrated her secrets'. The average man, then, as much as the scholar, the educated middle classes, the man of business and the man in the business of government, were well aware of the power of 'man' over nature. They were aware and proud of it, and celebrated this 'man', occasionally by staging monumental public rituals and spectacles.

The middle of the nineteenth century, with the industrial revolution in full swing, marked one of the best known public celebrations of the image of 'man' as the master of nature. In 1851 Queen Victoria opened 'The Great Exhibition of the Work of Industry of all Nations' in the Crystal Palace in London. As the naming of the event suggests, the exhibition was presented from the outset as a manifestation of global solidarity, tangible proof of the idea that 'the interests of all [nations] are closely interlocked' (Bury 1932: 331). Indeed, Prince Albert who had helped initiate the exhibition explained it as nothing less than 'the realisation of the unity of mankind'.[12] That this 'unity' was fragmented in practice by the division of the world and the allocation of different parts to the European colonial powers does not seem to have troubled the Prince. Being such a monumental event, the exhibition was no doubt polysemic, meant to convey a range of meanings and at the same time open to different kinds of interpretation. For one thing, as is to be expected, the British used the occasion to celebrate themselves by emphasising their dominant position in relation to other European nations and the rest of the world. As one commentator put it, 'we are ... and for the indefinite term continue to be, the great manufacturing and mercantile nation of the world'.[13] On a broader canvas, the exhibition was 'a public recognition of the material progress of the age and the growing power of man over the physical world' (Bury 1932: 329). Yet these grand abstractions aside, it was obvious to all concerned that the age of material progress that the exhibition celebrated did not belong to everyone, at least not in the same measure. Nor was the 'man' who had unlimited power over nature every man. As Stocking (1987: 3) points out, 'the most obvious lesson of the exhibition ... was that ... not all men had advanced at the same pace, or arrived at the same point'. This lesson was hardly missed by anyone.

In a series of lectures on the results of the exhibition, the Reverend Dr. William Whewell, Master of Trinity College, Cambridge, pointed out, using the terminology that was to become the staple of evolutionary anthropology, that the exhibition afforded a glimpse into the history and evolution of civilisation, the very order of the world itself. Under the same roof, the public could observe 'the infancy of nations, their youth, their middle-age, and their maturity'.[14] The Reverend did not need to assign specific nations to the different stages of evolution towards 'maturity'. If it was not obvious to the general public which nations occupied which position on the basis of the exhibits themselves – significantly enough, 'contributions from the British colonies and dependencies were for the most part in the Raw Material category' (Stocking 1987: 2)[15] – the commentary of politicians and writers during and after the exhibition left no room for doubt. In his study of the exhibition, Auerbach (1999: 197) summarises the gist of the commentary in the following way: 'In general, northern Europeans were held in the highest regard, followed by southern Europeans, with Russians, Asians, Africans, and American Indians bringing up the rear'. The Great Exhibition of 1851, then, was not merely a celebration of European 'man's' mastery of nature but also, as much, a celebration of European 'man's' mastery of humanity, which is to say, not only the attainment of what this 'man' posited as the essence of humanity but also mastery over all those Other men in the world beyond northern Europe who had not yet actualised this essence.

Victorian anthropology itself both reflected and lent scientific gravity to these images. As Stocking points out, by the 1860s the views of the leading Victorian anthropologists crystallised into a more or less general consensus, the major underlying assumptions of which were:

> That the motive force of sociocultural development is to be found in the interaction of ... [men] and the conditions of external environment; that the cumulative results of this interaction in different environments are manifest in the differential development of various human groups; that these results can be measured, using the extent of control over external nature as the *primary* criterion; that other sociocultural phenomena tend to develop in correlation with scientific progress; that in these terms human groups can be objectively ordered in a hierarchical fashion; that certain contemporary societies therefore approximate the various earlier stages of human development. (1987: 170; my emphasis)

These views found paradigmatic expression in the work of the best-known Victorian anthropologist, E.B. Tylor. As is well known, in *Primitive Culture*, Tylor constructs a three-stage evolutionary schema in which 'savagery' represents the lowest stage of cultural development, 'barbarism' the middle stage and 'civilisation' the third and highest stage. The criteria used in the construction of this schema were first, 'the industrial arts', second, 'scientific knowledge' and third 'moral principles [and] religious belief'. On the basis of these criteria, Tylor argued, the ethnographer was able to rank accurately existing societies, from lowest to highest in the development of civilisation. For example, 'few would dispute that the following races are arranged rightly in order of culture: Australian, Tahitian, Aztec, Chinese, Italian'. Although Tylor does not elaborate, one presumes that he meant 'Australians' and Tahitians to be occupying the 'savage stage', Aztecs and the Chinese to be in the middle at the stage of 'barbarism' and Italians and other Europeans to be at the top, the stage of 'civilisation'. Tylor recognised that his schema was more of an ideal-type construct, but nonetheless not very far removed from reality. Indeed, in the case of 'material and intellectual culture', the gap between the ideal and the real was particularly small. As he pointed out, 'acquaintance with the physical laws of the world, and the accompanying power of adapting nature to man's own ends, are, on the whole, lowest among savages, mean among barbarians, and highest among modern educated nations'. Moreover, even in the case of such complex and difficult to quantify elements of culture like morality, it was clear that civilised man was still far ahead of everyone else: 'Savage moral standards are real enough, but they are far looser and weaker than ours' (Tylor 1874: 26–27).

All in all, then, for Tylor, other Victorian anthropologists and most of their contemporaries, the evidence was overwhelmingly in favour of 'civilised man'. As Tylor argued, it goes 'far to justify the view that on the whole the civilised man is not only wiser and more capable than the savage, but also better and happier, and that the barbarian stands between' (1874: 31). And all this – wisdom, excellence, happiness – because 'civilised man' was able to do what Other 'men' could not, namely, master nature. If 'civilised man' had no rivers, he did not despair. He simply made canals. Such was the extent of his power.

'Europeans are devotees of power'

It is difficult today to understand the nonchalance with which European 'man' proclaimed his power over the physical world. It appears as sheer arrogance – indeed, for many environmentalists, nothing less than hubris. It is even more difficult to understand the uninhibited manner in which he celebrated himself and, reciprocally, the ease with which he denigrated all those 'men' who were different from him. To the contemporary sensibility, it can be nothing more than a blatant manifestation of deep-seated ethnocentrism. But if that is so, what is one to make of the fact that those who were denigrated for being different eventually came to see the world through the eyes of this 'man', recognised the truth of his pronouncements and in the process recognised themselves also as the sort of men that European 'man' said they were, namely, in fundamental respects inferior to him? What is one to say about the fact that they adopted his rhetoric, ideology, posture and sought to become like him, only to end up becoming a semblance of European 'man' – similar but never quite the same? Can we understand this fateful turn of events any better than European 'man's' arrogance and ethnocentrism? Moreover, can we afford to remember it any less?

How the modernist vision of reality managed to take firm hold of the non-European imagination and to colonise native consciousness is a question that cannot be settled easily. This is partly because of the sheer complexity involved. No doubt, there have been direct impositions on native consciousness, specific policies, institutional discourses and practices, material and symbolic technologies intended to 'civilise' the colonised. There have also been policies and practices with similar results even if not necessarily similar aims. Then there is the undirected, unplanned and unintended impact of travellers, men of business and settlers to be taken into account, their circulation among native populations and the circulation through them of European commodities, technologies and ideas. There have been accidental and fateful encounters, 'structures of the conjuncture', as Sahlins (1985) says, which often involved cultural misunderstandings, initiated exchanges of different sorts, generated spontaneous acts of defiance and led to unconscious incorporations of European assumptions by native populations. There were, finally, deliberate endorsements of the European 'man's' vision of the world and, no doubt, deliberate acts of resistance both of which developed over time with the benefit of, among other things, hindsight.

Beyond sheer complexity, however, there is another, even more intractable problem that plagues attempts to explain the colonisation of native consciousness. Any explanation must necessarily remain incomplete and therefore suspect. To explain why people think and act in certain ways and not others, anthropologists require a certain theory of practice, say, the insight that people are the product of their circumstances. If anthropologists are themselves the product of circumstances, however, how can they ever have access to such an explanatory insight? How can they know that circumstances make people when they themselves are people made by circumstances? The usual and no doubt arbitrary way of dealing with this problem is not to thematise the theoretical insight but to treat it as an exception to the rule. Hence, one ends up with a theory that can explain everyone else's practices but cannot explain itself. The point in raising this issue is not to criticise explanations of the colonisation of native consciousness for this deficit. All discourses that generalise on the human condition run into the same paradox and all are suspect in this sense. My aim rather is to draw attention to the fact that this deficit does not cease to exist because it is ignored. On the contrary, it persists and comes to bear heavily on the next step in the analytical process, namely, when one finally comes to the question of the de-colonisation of native consciousness, which is my main concern here. There have been numerous suggestions as to how native consciousness may be liberated but practice does not quite bear out the theories. If anything, those who do make the effort find themselves more entangled and securely trapped. As I will argue in detail below, this failure is not unrelated to the paradox that plagues the theorisation of how native consciousness is colonised. For the moment, however, it may be pertinent to turn to one of the more recent and influential anthropological theorisations of this sort, and to some of the more pronounced manifestations of the colonised consciousness.

There is by now a well-established tradition in anthropology known as 'historical ethnography' much of which has developed in response to what is often described as the economistic and mechanistic explanations of European domination of native populations. Although there are different strands to this critique, in broad terms the argument is that such explanations reduce native populations to passive objects upon which European practices and ideas were imposed, and colonialism itself to an imperialist quest for nothing more than profit and power. The critique operates within the same problematic as the earlier Marxist critique of the more orthodox version of Marxism and draws heavily on the notion of hegemony to counter arguments about the economy of self-interest and rational calculation and the politics of ideological manipulation. As the Comaroffs argue about the Tswana of southern Africa, the explicit messages of European culture, such as the word of God propagated by the missionaries, were for the most part rejected by native populations. Far more decisive for their 'conversion' was the introduction of the underlying forms of European culture into native lives, structures such as commodity exchange and the corresponding division of labour, language, science and technology which led to the internalisation of the terms of reference through which European culture could be endorsed (or rejected for that matter).

> Here, once again, lies the point. In being drawn into ... conversation [with the missionaries] the Southern Tswana had no alternative but to be inducted, unwittingly and often unwillingly, into the *forms* of European discourse. To argue over who was the legitimate rainmaker or where the water came from, for instance, was to be seduced into the modes of rational debate, positive knowledge, and empirical reason at the core of bourgeois culture. The Tswana might not have been persuaded by the substance of the claims made by the churchmen, and their world was not simply taken over by European discursive styles. Yet they could not avoid internalizing the terms through which they were being challenged. (Comaroff and Comaroff 1991: 213)

'The point', then, according to this explanation, was not the content of European culture – its discursive style, ethos, beliefs, values and so on – but its underlying forms. Native consciousness was colonised not so much from above, at the level of reflexive argumentation, demonstration and persuasion, as from below, the level of tacit incorporation by local populations of European assumptions about the nature of reality. The argument, in short, is that, much like the colonisation of the European working-class consciousness by bourgeoisie culture, this was not so much an ideological as a hegemonic process.

It seems that the power of the underlying forms of European culture to transform the lives of native populations was not lost on colonial officials and functionaries themselves. To turn to one characteristic example: the railway – which no doubt also proclaimed the 'Europeans' mastery of time and space' (Adas 1989: 224) and functioned as a symbol of European power – was considered among the most important mechanisms for social change in the colonies. By mastering the time and space of the colonised, it also helped to undermine traditional ways of seeing and being in the world and encouraged the development of new ones. As Adas (1989: 224–25) points out, colonial officials in India hoped that the introduction of the railway would accelerate the breakdown of the caste system: 'If high-caste Brahmins wished to travel by rail, they would have to rub elbows with low-caste farmers and laborers, thus making a shambles of notions about pollution through touch and encouraging social intercourse between members of different caste groups'. The expectation, then, was that by means of such underlying changes the local populations would come to see by themselves through personal experience the values and the value of European culture and amend their ways accordingly. Let us consider briefly some of the lessons that W.A. Rogers, an officer in the Indian Civil Service, believed the local population could learn through the introduction of the railway:

> Railways are opening the eyes of the people … in a variety of ways. They teach them that time is worth money, and induce them to economise; they teach them that speed attained is time, and therefore money, saved or made. They show them that others can produce better crops or finer works of art than themselves, and set them thinking. … They introduce them to men of other ideas, and prove to them that there is much to be learnt beyond the narrow limits of their little town or village. … Above all, they induce them in the habits of self-dependence, causing them to act for themselves promptly and not lean on others.[16]

There was, then, a whole host of values that Indians could be taught by the railway: the values of time, money, competition and efficiency and therefore the value of the capitalist economy itself, but also the values on which the capitalist economy is based – rational thinking, cosmopolitanism and universalism, independence and self-reliance.

It seems that W.A. Rogers was not wrong in his estimation. Many Indians did learn these lessons, even if not exclusively through the introduction of the railway. Indeed, some learned the most important lesson of all, the value of European learning. Take, for instance, Raja Rammohan Roy's petition to Governor-General Lord Amherst in 1823. At the time, the colonial government confined its support to education in India to the traditional forms of instruction, such as Sanskrit, Arabic and Persian. Roy, 'who was among the boldest of the early Indian leaders favoring sweeping reforms of Indian society and extensive Anglicization' (Adas 1989: 278), was concerned that this limitation would keep the country in the 'dark'. In his petition, therefore, Roy appealed to the Governor-General 'to promote a more liberal and more enlightened system of instruction, embracing Mathematics, Natural Philosophy, Chemistry, Anatomy and other useful sciences … which the nations of Europe have carried to a degree of perfection that has raised them above the inhabitants of other parts of the world'.[17] An education, then, not based on the sacred Sanskrit and Islamic texts or at least not exclusively. There was a fundamental need for Indians to be exposed to the more 'enlightened', utilitarian forms of education, the European sciences, which dealt with the empirical world, the here and now rather than the beyond, the sort of education that raised Europeans above everyone else in the world.

Such was also the message of another important nineteenth-century Indian figure, the nationalist novelist and satirist Bankimchandra Chattopadhyay. In an essay on Indian philosophy, which in many ways anticipates Max Weber's understanding of the connection between religion and material progress, Chattopadhyay singles out other-worldliness as the chief cause of the country's problems. 'The present state of the Hindus is a product of this excessive other-worldliness. The lack of devotion to work which foreigners point out as our chief characteristic is only a manifestation of this quality'.[18] Indians were not only lazy, according to Chattopadhyay, but also fatalists, and this trait too, no doubt also pointed out by foreigners, was the product of 'excessive' other-worldliness. 'Our second most important characteristic – fatalism – is yet another form of this other-worldliness derived from the Sāṅkhya [a system of Indian philosophy]'. 'Excessive' other-worldliness was responsible not only for the fact that India had become subject first to the Muslims and then to the British but also the fact that 'social progress in this country slowed down a long time ago and finally stopped completely'.

What India required to progress, Chattopadhyay pointed out, echoing Roy's appeal to Lord Amherst, was the sort of knowledge that Europeans produced, a knowledge that could be used to conquer life in the here and now rather than to win the life that lies beyond:

> 'Knowledge is power': that is the slogan of Western civilisation. 'Knowledge is salvation' is the slogan of the Hindu civilisation. … Europeans are devotees of power. That is the key to their advancement. We are negligent towards power: that is the key to our downfall. Europeans pursue a goal which they must reach in this world: they are victorious on earth. We pursue a goal which lies in the world beyond, which is why we have failed to win on earth. Whether we will win in the life beyond is a question on which there are differences of opinion.

Indians may or may not win the life of the beyond – no one can really tell. There was little doubt, however, that the sort of knowledge they traditionally pursued was not likely to win them the life of the here and now. To truly save themselves and advance like Europeans, they required a different kind of knowledge – the knowledge that unlocked the secrets of nature. To possess this sort of knowledge was to gain mastery of the earth, to bring it under one's control and bend it to one's wishes. Hence, the fact that Europeans became 'victorious on earth', which is to say victorious not only over the physical world but also over other peoples and nations. The colonisation of India and much of the rest of the world was one of the most apparent and tangible proofs of European power and this-worldly victory.

Although the situation was rather different in Africa, the educated élite was as convinced of the superiority of European culture and the inferiority of their own as educated Indians. A large number of these men were 'transatlantics', black Americans and West Indians who, as Davidson (1978: 172) points out, came to Africa running away from 'a white civilisation which would not have them but which they deeply admired' and sought to transform 'a black civilisation which would not have them either but which they in any case despised'. But there were also many indigenous Africans, mainly in western and southern Africa, who held similar views as the 'transatlantics' themselves. These were often the children and grandchildren of the so-called 'recaptives', enslaved west Africans shipped to the Americas but rescued by the British navy before reaching the New World and then 'put ashore as free people at Freetown in Sierra Leone' (Davidson 1978: 167).

One of the better known among such men was James Africanus Horton who trained in medicine at the University of Edinburgh and returned to West Africa as a Staff-Assistant Surgeon of the British Army. As July (1968: 114) points out, Horton, 'like most nineteenth-century Africans educated on western principles', believed that Africans should 'master the philosophy and techniques of ... European civilisation' since this civilisation was clearly 'superior to anything which had been produced by indigenous African societies'. Much like the Indian élite, Horton considered European education as the only way to escape 'ignorance and poverty, superstition and disease'. He was convinced that Africans could rise to the level of European civilisation 'by progressive advancement ... with a guarantee of the civilization of the north' (July 1968: 116). This guarantee was none other than European education. Accordingly, Horton proposed a system of education based on a curriculum that consisted of English grammar, arithmetic and geography at the primary level and, at the secondary level, in addition to these subjects, Latin and Greek, geometry, botany, mineralogy and music. He also proposed the establishment of a university in Sierra Leone where special attention was to be devoted to the physical sciences and mathematics, since these sciences were known to 'cure many defects in the wit and intellectual faculties'. More important, however, were the practical applications of these sciences. As Horton pointed out, 'Africa is considered as one of the richest mineral-producing countries [sic] in the world'. Moreover, 'the fields and forests of Africa abound with vast commercial and medical plants'. There was, then, enough natural wealth in Africa but not enough knowledge to take advantage of it. Luckily, Africans could turn to European science and technology for assistance. For was it not by these means that Europeans exploited the wealth of nature to achieve greatness? It was, indeed. 'Do we not find', Horton asked rhetorically, 'that the greatness of civilised countries, depends on the development and practical application of these studies[?]' If that is what we find, 'to whom then must Africa look for the ultimate development of her vast resources[?]'[19] It should no doubt look to the 'civilised countries' of the north which had made extensive use both of their own and other peoples' natural resources, achieved 'greatness' and could now afford to act as instructors and guarantors of Other nations' success.

It should not be assumed that endorsement of the modernist 'physics' and 'anthropology' by native élites was ever a matter of a wholly unqualified, once-and-for-all acceptance of European culture. The dynamics of symbolic domination are such that the same way in which rejection of European culture necessitates prior endorsement of its underlying forms, endorsement of its underlying forms often necessitates subsequent rejection of aspects of the culture itself. Acceptance by native élites of the ideology of progress involved a fundamental contradiction. It was both a recognition of European superiority and an affirmation, however grudgingly, of the inferiority of one's own culture. A nation without its own distinctive identity, however, without an essential character that made it what it is, could not become what it aspired to be – a nation among nations. Because nationalism set this fundamental limit on the would-be postcolonial state, European culture was to be accepted and rejected at the same time. Paradoxically, rejection became the very condition of possibility of acceptance. To illustrate this convoluted and paradoxical situation, consider a couple of examples. As Chatterjee (1993: 6) shows for India, European excellence was said to lie in the 'domain of the "outside", of the economy and of statecraft, of science and technology'. In this domain, European superiority was readily 'acknowledged and its accomplishments carefully studied and replicated'. Yet culture also consisted of an 'inner' domain, the universe of the spiritual, and here 'the East was superior' (Chatterjee 1986: 66). About this domain, India had nothing to learn Europe; on the contrary, it had things to teach it. Much the same strategy was adopted in Africa. As July points out (1968: 475), West African leaders such as Nkruma and Senghor had a vision of the postcolonial nation's historical mission in which Europe was expected to 'contribute her scientific knowledge' while Africa was to provide 'her spiritual strength and her realization that in art is found the meaning of life'. In the case of Africa too, the condition of possibility of acknowledging European superiority was its partial rejection.

It should not be assumed either that endorsement of the European vision of reality by native élites was ever a monolithic process that included everyone across the board. As we have seen, the modernist vision itself, although dominant, was by no means the only one in circulation. In the same way that many Europeans, most notably the Romantics, rejected it in favour of other, dominated visions, so did many natives. Gandhi is perhaps the paradigmatic figure in this respect, the non-European anti-modernist *par excellence*. In what is probably his most trenchant and systematic critiques of modernity, the *Hind Swaraj* (Indian Home Rule), European civilisation is depicted as 'civilisation only in name'. It claimed to be civilisation because it had achieved mastery of nature and had accumulated immense material wealth. Yet for Gandhi it was nothing of the sort. 'Formerly', he wrote, 'in Europe people ploughed their lands mainly by manual labour. Now, one man can plough a vast track by means of steam engines and can thus amass great wealth. This is called a sign of civilisation'. In the past, he goes to say in the same vein, 'men travelled in wagons. Now, they fly through the air in trains at a rate of four hundred and more miles per day. This is considered the height of civilisation'. Machines are now everywhere and the time will come when people would not need to do anything by themselves. But was this mechanisation that made 'bodily welfare the object of life' civilisation? For Gandhi, it clearly was not. There were far more important things in life than bodily welfare to which this 'civilisation in name' paid little or no attention – the welfare of the soul, morality and religion. It was the development of these aspects of human life, according to Gandhi, that accounted for true civilisation. In an ironic reversal of European accusations of lack of religion in native societies, Gandhi noted that European culture was characterised by its disregard for the religious life and that 'people in Europe … appear to be half mad' because of it – rich in material things, no doubt, but with disturbed minds nonetheless. This was by no means 'an incurable disease', Gandhi goes on to tell his compatriots, 'but it should never be forgotten that the English people are at present afflicted by it' (1963 [1910]: 18–21).

To say, then, that the European vision of reality colonised the native consciousness is to deny neither the complexity and paradoxical nature of this process nor to ignore the existence or importance of anti-modernist visions, such as Gandhi's. On the contrary, it is to underscore and emphasise both. It is also to raise an inevitable, fundamental and no doubt difficult question – the question, that is, of the de-colonisation of native consciousness. As I have already suggested, the strategies that may be used in such a quest ultimately depend on how the colonisation of consciousness is theorised to begin with. In certain anthropological and other circles concerned with what has come to be known as the 'politics of identity', this theorisation draws heavily on the reformed Marxist tradition, particularly the notion of hegemony. Here I shall examine briefly the argument developed by Jean and John Comaroff in their work on the Tswana of southern Africa that runs along such lines.

> We take hegemony to refer to that order of signs and practices, relations and distinctions, images and epistemologies – drawn from a historically situated cultural field – that come to be taken-for-granted as the natural and received shape of the world and everything that inhabits it. It consists, to paraphrase Bourdieu … of things that go without saying because, being axiomatic, come without saying; things that, being presumptively shared, are not normally the subject of explication and argument. … This is why its power has so often been seen to lie in what it silences, what it prevents people from thinking and saying, what it puts beyond the limits of the rational and the credible. (1991: 23)

The colonisation of consciousness, then, depends on the existence of a certain unthought, that which appears natural and necessary and therefore remains undiscussed and unquestioned. From here, the next logical step would be to suggest that de-colonisation must begin with the process of thinking this unthought, of forcing what goes without saying to speak and explain itself. When the unthought becomes an object of reflection, the Comaroffs argue, hegemony is no longer itself. It becomes 'ideology' and 'counter-ideology' – or in Bourdieu's (1977) terms, 'doxa' becomes 'orthodoxy' and 'heterodoxy' – and the social world the ground of argument and counter-argument, contest and struggle. This, for instance, is what has happened 'in the societies of the modern West' with respect to two other forms of domination, those based on gender and race. 'Formerly taken-for-granted discriminatory usages have been thrust before the public eye. As a result, the premises of racial and sexual inequality are no longer acceptable' (Comaroff and Comaroff 1991: 27). Is this, then, the way to go? Is this what natives should also do – think the unthought, develop a counter-ideology and resist the powers that be on their own ground, as Gandhi has done? If so, and this is the most critical question of all, does this amount to de-colonisation of their consciousness and the beginning of freedom? Although the Comaroffs clearly advocate reflection and resistance, they stop short of drawing such a conclusion. Notice the careful and equivocal phrasing of the argument: hegemony becomes 'ideology' and 'counter-ideology' – but is this to say that it is abolished, or that it changes form? The outcome of reflection and resistance to gender and race inequality is that their 'premises' are no longer 'acceptable' – but if so, how is the 'unacceptable' still reproduced? The Comaroffs provide no answers to these questions. As there may be good reasons for such evasiveness, it is important to examine the model of hegemony more carefully.

The first thing to note is the claim that in their encounter with the dominant culture, people take certain things for granted and that it is this unthought that constitutes the basis of their domination. In contesting the missionaries' explanation of rain, for instance, the Tswana unwittingly incorporated the underlying forms of European culture, a process that ultimately led to the colonisation of their consciousness: 'to argue over who was the legitimate rainmaker or where the water came from … was to be seduced into the modes of rational debate, positive knowledge, and empirical reason at the core of bourgeois culture' (Comaroff and Comaroff 1989: 213). The model also claims that when the dominated think the unthought and begin to question it, hegemony breaks down. In contesting the underlying assumptions of gender inequality, women opened up a domain of struggle against patriarchal domination and male privilege. These are two different moments in the unfolding of the model but bringing them together throws into relief the model's own unthought, which is also the problem that plagues it from the beginning. It should be apparent that at the heart of the model there is a fundamental contradiction. What the model claims in effect is that resistance against the dominant is both the condition of possibility of hegemony and a manifestation of its dissolution. Tswana resistance to the missionaries' message was the beginning of colonial hegemony, women's resistance to male domination the end of patriarchal hegemony. If one begins with a hegemonic situation in place – the stories of 'first encounters' notwithstanding, this is where one always begins – the contradiction signifies that, paradoxically, thinking the unthought generates a new unconscious domain, that resistance can occur only at the expense of reproducing hegemony further down the road. Bourdieu (1984), whose work the Comaroffs cite approvingly, seems to have been aware of this paradox. As he points out, because the dominant and the dominated agree on the stakes involved in the struggle – indeed, must agree if there is to be a struggle – it is immaterial whether one chooses to emphasise the disagreement that divides them in complicity or the complicity that unites them in disagreement. I shall return to this paradox and examine its structural conditions of possibility – which is to say also the conditions of the impossibility of modernist freedom – in the last chapter. For the moment, it may be useful to provide a concrete example from colonial history.

In his discussion of Indian nationalist thought, Chatterjee (1986: 51) calls the Gandhian revolution the 'moment of manoeuvre' and goes on to show that at this time the 'compromises' made during the earlier 'moments' in the nationalist struggle are denounced. As we have seen, nineteenth-century nationalists accepted European superiority in the external, material domain but insisted on Indian superiority in the internal, spiritual one. Gandhi, by contrast, rejected European civilisation writ large and completed the process of re-inventing and valorising a local tradition, the existence of which was necessary for the nation to become aware of itself as a nation. As Chatterjee points out, Gandhi's radical posture was dictated by the imperatives of the nationalist struggle and the politics of mobilisation. The 'moment of manoeuvre' was the time to unify the nation and prepare it for the actual struggle against colonial rule. Despite his complete rejection of European modernity, however, there was at least one thing that Gandhi could not reject – the idea of the nation. Without endorsing this idea and, as Chatterjee (1986: 112) points out, to a lesser or greater extent, the categories of 'post-Enlightenment' thought associated with the idea of the nation, Gandhi would have no case against British colonial rule.[20] The nation was the unthought that Gandhi took for granted in order to think the unthought and to construct an ideology that would counter colonialism. It is the complicity that united Gandhi and the British in disagreement – or the disagreement that divided them in complicity – the hegemonic unwittingly taken on board in order to do away with the hegemony that had colonised Indian consciousness. Such is the paradox with symbolic domination.

> The major theoretical problem is to distinguish between alternative and oppositional initiatives and contributions which are made within or against a specific hegemony (which then sets certain limits to them or which can succeed in neutralizing, changing or actually incorporating them) and other kinds of initiative and contribution which are irreducible to the terms of the original or the adaptive hegemony ['ideology' or 'orthodoxy'], and are in that sense independent. (Williams 1977: 114)

The theoretical problem is far more complex and intractable than Williams suggests. If the aim is to do away with a 'specific' hegemony, there can be no 'outside' to it and no 'not-against' it. One is already within, against and dependent on it by virtue of the aim itself. To become aware of a specific hegemony as hegemony is already to take on board its most fundamental assumptions.

As it turned out, the paradox with hegemony reproduced itself in a less convoluted way. In order to resist European civilisation at all, Gandhi could do nothing else than to find a firm foothold within it – the nation. Unlike him, the men who led the new postcolonial nations – men such as Nehru of India, Soekarno of Indonesia, Nyerere of Tanzania, Nkrumah of Ghana, Azikiwe of Nigeria – found more than just a foothold. They too denied 'the alleged inferiority of the colonised people'. But they 'also asserted that a backward nation could "modernise" itself while retaining its cultural identity' (Chatterjee 1986: 30). As Nehru characteristically put it in a letter to Gandhi who was raising questions about the wisdom of embarking on a large-scale industrialisation, 'we are trying to catch up, as far as we can, with the Industrial Revolution that occurred long ago in Western countries'.[21] Unlike Gandhi who eventually abandoned politics, these men stayed on and in trying to 'catch up', continued to contest the inferiority of their culture only to reproduce it in another form – as 'modern' nations but always not quite yet. To this last phase of the triumph of the European vision of the world – of the modernist 'physics' and 'anthropology' – which is also the period of colonial independence and postcolonial dependence, I turn next.

'The Leap across the Centuries'

The end of the Second World War brought a shift in the paradigm in which the Western vision of reality was couched. While European confidence in its vision was declining, partly because Empire itself was crumbling and partly because of the experience of the two wars and the destructiveness they had unleashed, American confidence in its own version of this vision was gaining strength. The new paradigm to emerge out of the United States came to be known as 'modernisation and development' and was soon to replace the European civilising mission as the pre-eminent ideology of Western dominance.[22] With the new paradigm in place, the hierarchy of the world assumed a different guise. Gone was the colonial differentiation between civilised, barbarous and savage peoples. The lines of division were now drawn between developed and underdeveloped nations ('less developed' or 'developing' in subsequent, more cross-culturally sensitive literature), modern and traditional societies, the First and Third World.

Despite changes in terminology, and the differences between the two paradigms notwithstanding – for one thing, the American paradigm was far more systematically developed so that, for example, development and underdevelopment could be quantified and measured precisely – the vision of the world depicted in the American model remained fundamentally the same. As in the European paradigm, the measure of cultural value and worth among peoples and nations was to be found in 'man's' mastery of nature made possible by the power that science and technology put in his hands. 'What we envisage', Harry Truman announced in his inaugural address as the new president of the United Sates in 1949, 'is a program of development. ... Greater production is the key to prosperity and peace. And the key to greater production is a wider and more vigorous application of modern scientific and technical knowledge'.[23] Two years later the experts commissioned by the United Nations to suggest 'measures for the economic development of under-developed countries' articulated this vision in even more explicit and uncompromising terms. 'Progress', they proclaimed, 'occurs only when people believe that man can, by conscious effort, master nature' (United Nations 1951: 13). That was the fundamental lesson to be learnt by those who had not yet learnt it. If the 'underdeveloped' nations of the world wished to progress, they first had to recognise that nature was 'masterable'. From then on, it was largely a question of a 'rigorous' application of science and technology. Nature would then oblige and bend to 'man's' will, even the will of the 'underdeveloped man'.

There were, of course, many preconditions to development and progress. To begin with, the United Nations experts pointed out, 'the government of an under-developed country should ... make clear to its people its willingness to take rigorous action to remove the obstacles to free and equal opportunity which blunt the incentives and discourage the efforts of its people' (UN 1951: 93). Then there was a whole host of other, equally important issues to be considered: economic organisation and planning, domestic capital formation, credit facilities, transfer of technology from the developed countries, the development of a basic infrastructure such as roads, ports and airports, issues of land reform and employment, education, health and population growth – all in all, a formidable and daunting task even for the most determined and capable government of an 'underdeveloped' country. Of all the preconditions however, the most fundamental was the inculcating of the 'underdeveloped man' with the belief in the unlimited powers of 'man' – with the kind of humanism, that is, which three decades later was to be condemned and denounced as arrogance and hubris. Nothing was impossible for 'man' once he set his mind to it, proclaimed confidently the European luminaries of the nineteenth century – as we have seen in Buckle, there was no evidence 'which authorises us to assign even an imaginary boundary at which the human intellect will, of necessity, be brought to a stand'. 'In the long term all things are possible', insisted the United Nations experts in 1963 in the context of the 'United Nations Conference on the Application of Science and Technology for the Benefit of the Less Developed Areas', urging the people of the less developed areas to grasp the opportunity in a 'World of Opportunity'. All things were still possible in the 1950s and 1960s, it would seem, including, the people of the 'less developed areas' of the world were informed, the 'unlimited expansion of the sphere of human inhabitancy' in space. Not that such a thing was strictly speaking necessary. As the same authorities insisted, 'planet Earth itself might support a population many times – perhaps even ten times – the present one'. Indeed, 'the question [was] not "How?", but "When?". The difficulties [were] in the short term' (UN 1963: 232–33).[24] The 'developed man', then, had no doubt about what 'man' could do in the long run with the use of science and technology. The trick was to get the 'underdeveloped man' to believe in the same thing and to act accordingly.

Such was the trick. 'Underdeveloped' nations were economically backward precisely because they were 'traditional'. They lived the past in the present and reproduced it over the centuries with little or no change. Take technology, for instance. It was striking, the United Nations experts pointed out, how low the level of technology was in the 'underdeveloped' countries. Striking perhaps but not unexpected. In these societies, technology, much like anything else, was reproduced exactly as it was handed down by earlier generations. 'In certain fields of production, some of these countries have made no improvement in technology for centuries'. How many centuries? In some cases, more than twenty. 'In some parts of the Middle East … agricultural techniques are no better today than they were in the times of Pharaohs. In fact, in the field of irrigation there has been definite retrogression' (UN 1951: 28). 'Underdeveloped' countries, then, would not develop economically unless they transformed themselves into 'modern' societies. Development and modernisation went hand-in-hand. Yet as the United Nations experts warned, modernisation was not likely to be an easy task. It required radical and often painful changes in the way that 'traditional man' thought about the world and acted in it. 'There is a sense in which rapid economic progress is impossible without painful adjustments. Ancient philosophies have to be scrapped; old social institutions have to disintegrate; bonds of caste, creed and race have to be burst' (UN 1951: 15). Nothing less than a complete cultural transformation would save 'traditional man' from economic backwardness, poverty and disease; nothing less than 'a radical change in the outlook of the peoples of the under-developed countries' would bring them in line with the countries of 'Western Europe and the United Sates of America' (UN 1951: 30).

Assuming, then, that 'traditional man' managed to escape his predicament, how would he look after he had operated on himself and removed all his deficiencies? Not surprisingly, he would look very much like the 'man' that 'modern man' imagined himself to be. Let us consider briefly the image that the American social scientists Inkeles and Smith sketched in the 1960s – an image quite influential at the time and which has been recently resurrected in fundamental respects in modernist social theory.[25] To begin with, 'traditional-cum-modern man' would be open to new experience and prepared for social changes at more or less any time; he would be willing and able to offer opinions on a wide range of topics and active in seeking to acquire information (because his opinions would be worthless without the backing of facts); his time orientation would be such that he would be prepared to defer immediate gratification for future rewards; he would have a sense of efficacy, which is to say, an awareness that he could make an impact, if he chose to do so, both on the natural and social world; he would be educated, optimistic about the future and have occupational aspirations; he would reject the narrow value system by which he judged himself and his fellow men and adopt universalistic standards, become politically involved, loosen his kinship ties, forget patriarchy and, last but not least, put God where he belonged – in the world beyond.[26] He would, in short, be the master of the world and of his own destiny.

The rhetoric of development was quickly seized by native élites and was incorporated in the demands for independence. As Chatterjee (1993) shows for India, by the 1940s the emphasis of these demands shifted from the illegitimacy of colonial rule as foreign political domination to its illegitimacy as an exploitative force that impeded the development of the nation and perpetuated a backward economy. Indeed, as we shall see below, this theme was to become one of the fundamental tenets of the Non-aligned Movement. The rhetoric of modernisation had also made significant inroads in the way that native élites saw and projected themselves and their nations. 'The Tanzanian people now know that our poverty, our ignorance, and our diseases, are not an inevitable part of the human condition', President Nyerere said in 1969 during a visit to Canada. 'Once we accepted these things as the will of God; now they are recognized as being within the control of man'. Hence, the President went on to say, 'political freedom is ... no longer enough for us' (1973: 110). A few months later, President Nyerere repeated the same humanist message to the Tanzanian people itself. On declaring 1970 as Adult Education Year he noted: 'We are poor, and backward and too many of us just accept our present conditions as "the will of God", and imagine that we can do nothing about them' (1973: 137). Apparently, adult education was meant to dispel such long-held misconceptions.

The Indian élite, as we have seen, had mastered a similar rhetoric well before independence. Although 'religions have helped greatly in the advance of humanity', wrote Nehru in *The Discovery of India*, recapitulating the arguments of the nineteenth-century Indian modernisers, they also caused a great deal of damage. 'Instead of encouraging curiosity and thought, they have preached a philosophy of submission to nature' – a philosophy premised on 'the belief in a supernatural agency which ordains everything'. Science, on the other hand, 'made the world jump forward with a leap, built a glittering civilization … and added to the power of man to such an extent that for the first time it was possible to conceive that man could triumph over and shape his physical environment'. Nehru was writing in the aftermath of the Second World War and could not but note also that although 'man' could master nature, he proved unable to master himself. Nonetheless, he goes on to say, not everything was lost, and science could still save the day. 'Perhaps new developments in biology, psychology, and similar sciences … may help man to understand and control himself more than he has done in the past'. Indeed, for Nehru, much like the European luminaries of the nineteenth century, there was 'no visible limit to the advance of science, if it is given the chance to advance' (1961: 543–44).

In India, plans to make the country 'jump forward with a leap' were in place well before independence, despite Gandhi's objections. 'The three fundamental requirements of India, if she is to develop industrially and otherwise', wrote Nehru in 1946, 'are: a heavy engineering and machine-making industry, scientific research institutes, and electric power. These must be the foundations of all planning, and the National Planning Committee laid the greatest emphasis on them'.[27] Nehru's determination to transform India from a 'backward, traditional' society into a 'modern' one found practical expression in the Soviet-style five-year plans which his government implemented after independence. The rationale behind the overall effort was clear. 'Unless a social group or country changes', he pointed out on the occasion of inaugurating a machine tool factory, 'it loses its pre-eminence and becomes backward'. Pre-eminence of the nation in the community of nations, then, was the driving force; and science and technology were the means through which the goal would be achieved. 'In modern life', Nehru goes on to say in the same speech,

> Science and the progeny of science, techniques, technology, etc., are of the highest importance. They govern our lives and the conditions of living today. Therefore, we should understand and profit from them. What is happening today behind the Five Year Plans and other economic programmes in India is the change-over from the traditional society into a modern society.[28]

The will to modernisation and development was shared by most, if not all the leaders of the newly independent nations. As Davidson (1978: 316–17) points out, in Africa 'the new ruling groups … had already accepted the political lessons of the capitalist model, notably those of parliamentary democracy on the pattern of Westminster or the Palais Bourbon. ... Now they were more than ready to accept the economic lessons'. Already in 1955 when much of the world was still under colonial rule, twenty-nine heads of state from Asia and Africa met in Bandung for what was effectively the first major conference on development organised by the 'underdeveloped' countries themselves. The conference, known as the 'Asian-African Conference', was the precursor of the Non-aligned Movement launched in Belgrade six years later and which was to champion the cause of development in the 1960s and 1970s.[29] In the 'Final Communiqué' of the Bandung conference, the participating states 'recognised the urgency of promoting economic development in the Asian-African region' (Jankowitsch and Sauvant 1978: lvii). And having realised the potential of the energy source that was discovered by military science and technology during the war, they 'emphasised the particular significance of the development of nuclear energy for peaceful purposes, for the Asian-African countries' (1978: lix). They also noted that there were 'many countries in Asia and Africa which have not yet been able to develop their educational, scientific and technical institutions' and recommended 'greater co-operation among them'. In a similar vein, the 'Declaration of the First Conference of Heads of State or Government of Non-aligned Countries' held in Belgrade in 1961 noted the 'tremendous progress [that] has been achieved in the development of science, techniques and in the means of economic development' (1978: 4). At the same time, they pointed out that there was an 'ever-widening gap in the standards of living between the few economically advanced countries and the many economically less-developed countries' and agreed that the gap must be closed as a matter of urgency 'through accelerated economic, industrial and agricultural development' (Jankowitsch and Sauvant 1978: 6).

This urgency prompted a 'Special Conference' in Cairo one year later which was attended mostly by technocrats from the various member-states and which was entirely devoted to the 'Problems of Economic Development'. The delegates noted once again 'the growing disparity in the standards of living prevailing in different parts of the world' and used the opportunity to complain 'that despite universal acknowledgement of the necessity to accelerate the peace [sic] of development in less developed countries, adequate means of a concrete and positive nature have not been adopted [to assist them]'. Of particular concern were also the terms of trade that 'continue to operate to the disadvantage of the developing countries'. Such unfavourable conditions notwithstanding, 'the developing countries have made progress in their economic development' and were determined to progress even further. There were 'new opportunities for ... cooperation among developing countries' in such fields as 'education, research, technical assistance, trade, industry, transport and communication', which could be exploited for mutual benefit. All in all, the message of the conference was optimistic and reflected the confidence of the 'less developed' countries in the unlimited powers of 'man'. The delegates 'affirm[ed] that the economic and social problems of developing countries could be solved *effectively* within a *reasonably short period of time*' (Jankowitsch and Sauvant 1978: 72; my emphases).

It was, then, a question of time for the 'developing' to transform themselves and achieve plenitude in their being, a full presence against the absences and deficiencies that both marked and limited their existence. It was only a matter of time, and a relatively short time at that. For was not the destiny of 'man' to master the world and bend it to his will? And was there anything stopping him apart from his self-induced ignorance and superstition? Were not 'man's' powers unlimited? Did he not have science and technology at his disposal? And had they not proven themselves over and over again by working miracles in every sphere of life? Was not progress, therefore, stretching into infinity beckoning? This humanistic message was forcefully reiterated once again a year after the Cairo Non-aligned conference in another conference on development, this time organised by the United Nations. 'Just as primitive man escaped from the limitations of his environment by clothing his nakedness in pelts and meeting the cold with fire', the delegates were told, 'so modern man has extended the frontiers of his environment to live in civilized terms in the coldest or the hottest climates'. There were, of course, many other examples of modern 'man's' civilised existence:

> [He] has made the air his highway. He has accommodated himself to the speeds and the pressures, and the non-pressures of space travel. He has travelled for months on end in nuclear submarines in the depths of the ocean. He has discovered the disease-causing germs which can now be avoided or overcome by serotherapy, vaccinations, chemotherapy, antibiotics, pure water supplies, the treatment of sewage, etc. He has identified the vitamins and the hormones and the principles of sound nutrition. He has carried out operations on the brain and the heart, and has maintained life by extra-corporeal circulation. (UN 1963: 34)

All these were indisputable facts, testimony to the powers of 'man' and his handmaidens – science and technology. True enough, the conference organisers noted, voicing environmental concern perhaps for the first time, 'by his very success he has created new problems [such as] the chemical contamination of the atmosphere, of the water supply and of foodstuffs' (1963: 34). But these were mistakes that need not be repeated and which, in any case, could be solved just as easily through the use of science and technology. Although 'warnings of a misguided use [of resources] in the past', these problems also provided 'insight into the hopeful opportunities now offered by modern science and technology' (1963: 10).

As I have already pointed out, the first volume of the proceedings of this particular United Nations conference – on the 'Application of Science and Technology for the Benefit of the Less Developed Areas' – is auspiciously entitled 'World of Opportunity'. Its key message to the 'developing' countries was that they should grasp the opportunity offered by the historical conjuncture and make the most of it. The argument was that the 'developing' countries that had just emerged from centuries of colonial domination were beginning their life as independent nations with a fundamental advantage. Unlike the 'developed' countries, they did not need to invent the wheel. It had already been invented and was ready at hand. Hence, they could use science and technology to 'leap across the centuries' to the Western present. This slogan, the report says, 'was used repeatedly during the conference' – and with good reason. It drove home the critical message that the 'developing' countries could achieve 'rapid transition from a primitive to a modern economy by vaulting the intermediate stages of development which the more advanced countries had already undergone' (UN 1963: 221). Why grope in the dark trying to find solutions to problems that have already been solved, repeat the same mistakes in the process, learn through trial and error, follow every single step in the difficult march towards progress and civilisation? Thanks to the 'more advanced' countries, the path was now wide open and could be traversed with a single, daring, long jump.

A happy historical conjuncture, then, and an auspicious message. Having learnt of his 'primitive' condition, his 'backwardness' and his 'underdeveloped' state; having come face-to-face with the 'poverty, ignorance and disease' that afflicted him for centuries; having realised that without becoming a 'devotee of power' he would remain forever subject to those who were devoted to it; having come to recognise that 'political freedom was not enough' to assure him dignity and respect in the world; having seen by himself the 'glittering civilisation' that lay across the centuries beckoning seductively; having, finally, had a glimpse of the enormity of the task lying ahead, the 'traditional man' paused for a minute – hours, years, decades even – to compose himself and reflect. Was there anything else to do but to find a 'foothold', brace himself and take the plunge? Should he not take advantage of the unique opportunity offered so magnanimously by his fellow man – the 'modern, advanced man' – to 'leap across the centuries' and land in a matter of decades on the other side of time? Should he hesitate further, procrastinate more, waste even more precious time?

The 'traditional man' did, of course, take the plunge in the end. He embarked on what promised to be the fastest journey in history but strangely enough, he has not arrived yet. He is still in mid air, hovering uncomfortably, looking in vain for a place to land. Something must have happened during transit; something must have gone terribly wrong on the other side of time. In the place 'across the centuries', time has all but stopped, is hesitating to proceed further and the 'man' who has been riding it for so long is hesitating to urge it on. For the first time in centuries, time is looking surreptitiously over its shoulder rather than steadfastly forward, is reconsidering its evolutionary certainties and one-way determination and is quickly filling up the present with its sheer volume. There is hardly any space left for anyone else, and the 'traditional man' is told to hold his position – and to hold on. Having taken the plunge across the centuries, it looks as though he is now trapped between them.

Notes

1. For an anthropological study of contemporary Greek perceptions of nature, see Theodossopoulos (2003). See also Argyrou (1997).
2. For an anthropological critique of Eurocentrism in relation to classical Greece, see Herzfeld (1987).
3. Anthropology must be one of the few Western discourses to admit to this failure. It recognises it in the critique of its own discourses as ethnocentric, whether this is directed against specific paradigms – evolutionism, for instance, or structuralism – or, as in the case of so-called 'postmodern anthropology', against the discipline writ large.
4. Much of the following discussion draws on Adas's excellent study of 'science, technology and ideologies of Western dominance'.
5. As well as, Bourdieu points out (1977, 1990), the neutralisation of time and the urgencies of everyday life.
6. Quoted in Hodgen (1964: 199; my emphasis).
7. Quoted in Morris (1987: 91).
8. Quoted in Pagden (1982: 78).
9. 'The *imagination* of man' wrote Hume (1977 [1748]: 112), 'is naturally sublime, delighted with whatever is remote and extraordinary, and running, without control, into the most distant parts of space and time in order to avoid the objects, which custom has rendered too familiar to it. A correct *judgement* observes a contrary method'.
10. Quoted in Stocking (1987: 153).
11. Quoted in Buckle (1878: 154).
12. Quoted in Bury (1932: 330).
13. Adam Ward quoted in Auerbach (1999: 166).
14. Quoted in Stocking (1987: 5–6).
15. The other major exhibit categories were Machinery, Manufactures and Fine Arts.
16. Quoted in Adas (1989: 226).
17. Quoted in Charles O. Trevelyan (1838: 66).
18. All Chattopadhyay quotations from Chatterjee (1986: 56–7).
19. Quoted in July (1968: 121–22).
20. As Chatterjee shows, although Gandhi's initial vision of the nation was largely outside 'post-Enlightenment' thought, as the struggle against the British developed, it became increasingly implicated in it.
21. Quoted in Chatterjee (1993: 202).
22. The literature on modernisation and development is immense but for early, classic statements see, in particular, Rostow (1960) and Inkeles and Smith (1974). For a recent, poststructuralist critique see Escobar (1995).
23. Quoted in Escobar (1995: 3).
24. The proceedings of this conference were published in eight volumes. The first volume, devoted to science and technology, is entitled 'World of Opportunity', and chapter seven, 'Grasping the Opportunity'.
25. See, in particular, Anthony Giddens (1991) and Urlich Beck (1992a). For a critique see Alexander (1996) and Argyrou (2003).

26. See Inkeles and Smith (1974).
27. Quoted in Morehouse (1969: 3).
28. Quoted in Morehouse (1969: 12).
29. As Jankowitsch and Sauvant (1978: xliii) point out, it was not until 1973 at the Algiers summit that 'development and related economic issues [emerged] as the principal objectives of the non-aligned movement'. Until that time the primary concern was the consolidation of political independence, even though 'concern with economic development had always been present – as expressed already at the Bandung conference and especially at Cairo [in] 1962'.

2 Second Change

The time came when nature – that entity which for centuries was perceived and treated as an object to be tamed, mastered and brought under 'man's' control – ceased to exist. And, eventually, so did its master. What replaced this nature in 'European thought' is a very different sort of being, another nature whose nature has very little to do with that of its predecessor. What replaced 'man' the master of nature is also a very different sort of being – a being whose status in the wider scheme of things has been fundamentally altered, if not completely reversed. To begin the process of unpacking this momentous change, and to avoid confusion, let us call the old nature 'nature' and the being that replaced 'man' by the name it has acquired – the 'human being'.

The death of 'nature' has given birth to what most people would now agree is a fragile domain, an entity hardly amenable to taming, compelling, mastering, or bending to 'man's' will. Nature, it is now recognised, is a system of immense complexity and delicate balance whose future is currently in the balance. It has been burdened with numerous afflictions – deforestation, acid rain, depletion of the ozone layer, the greenhouse effect, the contamination of the seas, lakes and rivers, the poisoning of the food chain, the extinction of species and consequent depletion of biodiversity. It is at the brink of collapse and must be protected and cared for as a matter of urgency. The death of 'man' has also given birth to a new creature, one that claims to be more perceptive and receptive, better placed to understand the order of things and its place in this order, a more cautious and considerate being. Gone are the manly postures of confrontation, the language of subjugation and exploitation, the images of 'glorious struggles'. Human beings, it is now pointed out, can no longer afford to be locked in combat with nature. They have caused enough, possibly irreversible damage. They must disengage, relent and make peace. They must relearn how to coexist and live in harmony with nature.

Such is the consensus in an otherwise complex and fragmented field – the environmentalist paradigm. To retain some of this complexity, initially at least, it is necessary to distinguish between two broad environmentalist tendencies. The first appears to be purely instrumental and functional but as I will argue in detail below, this is not necessarily the case. For the moment let us note that according to this view, we must save nature to save ourselves. The argument, of course, refers in the first instance to us in the here and now. Yet the ecosystem is not likely to collapse during our lifetime and even if it does, the full impact will be felt by future generations. Whether explicitly stated or not, then, the 'us' includes also those who will come after. It refers to humanity writ large and in the abstract, an 'imagined community' in Anderson's (1991) sense – which already suggests that this view may not be so instrumental and functional after all. The second, and more radical view is primarily ethical: we must save nature not only or even mainly for our sake but also and especially for its own – precisely because nature is not ours to destroy or do with as we wish. In the first view, human beings are 'stewards' of nature, in the second, they are simply beings among other beings – for some radical environmentalists worse than other beings, for most not better in any essential sense. Both sides point to environmental pollution and express the same degree of concern. The first uses the term literally, the second understands it, in addition, as defilement. For the first, nature is vital, for the second, it is vital and sacred.

The 'death of nature and the rise of [the] environment' (Escobar 1995: 192), which signifies at the same time and is in many ways the same thing as the death of 'man' and the birth of the 'human being', is without doubt an ontological transformation of the highest order. It is a radical change in the meaning of both the physical and human world, an event that heralds the emergence of a new 'physics' and a new 'anthropology' and, therefore, also a new order of things. What is one to make of such an epoch-breaking and -making event? How is one to understand it? What caused such a radical change? How did it come about? And what does it mean? No doubt, there are well-rehearsed, well-known and widely accepted answers to these questions. One could say, for instance, that after centuries of misconceptions and miscalculations, improvidence and arrogance, utilitarian rapacity and greed, the world has finally come to its senses. Stark reality forced itself upon us, drove home the implacable message of imminent ecological collapse and left humanity with no choice but to take immediate action. Yet there are problems with such an answer, chief among them the problem of reducing the issue to a perception of 'facts', as if the categories through which things are perceived and understood are neutral and transparent, as if environmentalists operate with immaculate conceptions that originate outside history and culture and are not affected by them. As we shall see below, environmentalists often wonder – and despair – about public apathy in the face of so many 'incontrovertible' environmental facts. Yet there is no mystery here. 'Facts' become visible, relevant and meaningful – which is to say facts – under determinate cultural conditions. It is these conditions that need to be explored if we are to gain a broader understanding of the ontological change that is currently upon us. It is the wider historical and cultural context we must sketch if we are not to remain hopelessly caught up in it, innocently attached – which is not to say innocuously – to untenable positions, whether environmentalist or counter-environmentalist.

I shall turn to this task in the next three chapters. My primary concern in the present chapter is the change itself. I shall explore this change, in the first instance, by examining some key United Nations documents that deal with economic development and the environment. Although the United Nations were relatively slow to react to the new 'physics' and 'anthropology', their documents are important in another sense. They represent the most authorised, institutionalised and widespread version of 'mainstream' environmentalism.

'Only One Earth'

In the early 1950s, it was still possible to speak about 'man' and 'nature' – still possible, that is, to speak about the meaning of the world in pretty much the same way as the luminaries of the nineteenth century. The experts could still argue with no hesitation or second thoughts that 'nature' was an object to be mastered and brought under 'man's' control and that this was an important historical lesson to be learnt by everyone, particularly the 'less developed' countries that had a different view of themselves and their surroundings. Let us recall the advice of the United Nations experts to the 'less developed':

> Economic progress will not be desired in a community where the people do not realise that progress is possible. Progress occurs where people believe that man can, by conscious effort, master nature. This is a lesson which the human mind has been a long time learning. Where it has been learnt, human beings are experimental in their attitude to material techniques, to social institutions, and so on. This experimental attitude is one of the preconditions for progress. (UN 1951: 13)

Such, then, was the universal message of progress, a message of pure positivity. 'Man' *can* – this was the fundamental lesson to be learnt. Through conscious effort, he can master nature, change society, transform himself, conquer the world. 'Man' can and *should* – for it was only to the extent that he did what he was capable of, that 'man' would at long last fulfil his destiny and become his true self.

As we have seen already, by the early 1960s the rhetoric, if not the substance of the humanist message was beginning to change somewhat. A new set of notions appeared in the lexicon of progress and development, words that spoke of 'improvidence' and recommended 'caution'. 'This is a small planet', began the Report on the United Nations Conference on the Application of Science and Technology for the Benefit of the Less Developed Areas (UN 1963: 3). It is a small planet that has been entrusted to 'man' but his 'stewardship has been improvident thus far'. In his eagerness to take control and assert himself, 'man' has forgotten how small and fragile the planet is and has been causing serious and possibly irreversible damage. Take deforestation, for instance. 'The fertility of the earth depends on nine inches of soil … [which makes it] the most important natural resource'. Yet the 'indiscriminate cutting down of the forests has meant that the rains, which were once tempered and filtered into the underground streams, now scour the soil from bereft hillsides'. The trees are gone, the soil is gone and so is the water from the rains that now finds its way to the sea and is thus wasted. Or take 'man's' interference with the atmosphere. 'By burning fuel, human activities affect the atmosphere. About 6,000 million tons of carbon dioxide are mixed with the atmosphere annually [which] can disturb the heat-balance of the earth, because of what is known as the "greenhouse effect"' (UN 1963: 7). Similarly, the Report goes on to say, 'human activities are interfering with the hydrosphere and the balance of nature which depends upon it'. The problem here – 'notoriously' – is the 'discharge of oil-wastes from tankers into the sea which, apart from the effects on sea creatures, has had disastrous effects on bird-life' – a lesson not restricted to the sea and the creatures that live off it but 'can be extended, in multifarious forms of pollution, to the abuse of the waters of rivers and lakes' (UN 1963: 7–8).

A sobering reminder, then, for the 'improvident steward', the 'modern developed man', and no doubt a message also for the 'less developed man' to heed – a sobering reminder about the paucity of the planet, the scarcity of resources, the fragility and precarious balance of nature. Yet the world of the early 1960s was still an optimistic world with its faith in science and technology more or less intact. Accordingly, and the warnings notwithstanding, the United Nations Report goes on to underscore the need for an 'injection of science' into the 'less developed societies' and calls on them 'to take science judiciously in their way of life' (UN 1963: 223). As everyone knew – and as many people still believe – science and technology can work miracles. Among other things, the Report says, they can 'sweeten the seas and harvest the deserts'. Indeed, the United Nations experts thought that this idea 'would be a good slogan for the Development Decade' of the 1960s (UN 1963: 242) – a decade that, significantly enough, was expected to combine and carry out two revolutions at once: 'the revolution of rising expectations and the revolution in science and technology, which can bring those expectations to life' (UN 1963: 243). A more upbeat and auspicious message, then, but one nonetheless delivered with a sour note already attached to it that diluted the promise of sweetened seas and blossoming deserts somewhat. All in all, the message of the United Nations Report was mixed and ambivalent, reflecting the contradictions that were for the first time beginning to emerge: faith in 'man' who turned out to be 'improvident' and not as thoughtful as he had imagined himself to be – the 'man' who was not afraid to attempt anything and was capable of everything, including causing irreparable damage to what he entirely depended upon. As such, it was also a message that could not but leave the 'less developed', by now thoroughly committed both ideologically and practically to accelerated economic development and progress – the 'leap across the centuries' – puzzled, apprehensive and suspicious about the newly acquired concern for nature. There was more to come.

In 1972, the United Nations organised in Stockholm its first conference on the 'Human Environment' with an eye on the question of the development of the 'developing countries'.[1] The first statement of the Declaration is indicative of the changes in the modernist vision of the world that had already taken place and sets the mood and tone for the rest of the text. 'Man', the Declaration begins, 'is both creature and moulder of his environment, which gives him physical sustenance and affords him the opportunity for intellectual, moral, social and spiritual growth'. Apparently, 'man' was still very much in evidence in the early 1970s but a more careful reading of the statement reveals that this is not quite the kind of 'man' that we have come to know so well. This 'man' has been drastically reduced in size. He no longer appears as the indisputable master of nature but more a humble 'moulder'; rather than the creator of the world, he now emerges as a 'creature' in it; and in case he still had an inflated image of himself, the Declaration reminds him of his manifold dependence on nature. What we encounter in this text, then, is not the figure of 'man' but the debut of a new and much more modest persona, the 'human being'. As for 'nature' itself, it is nowhere to be found. From the very beginning of the Declaration and already in the title of the Conference itself, 'nature' has been replaced by the 'human environment'. Rather than an intractable force to be tamed and a domain of utility to be exploited, the physical world emerges in its most positive light yet, bearing gifts and affording numerous possibilities. It now appears as the 'giving environment',[2] providing generously to human beings and taking care of their many and diverse needs, from their physical survival to their intellectual, spiritual, moral and social growth.

Having outlined the contours of the 'physics' and 'anthropology' that were to develop further in the following decades, the Declaration turns quickly to what has gone wrong to issue a sombre warning: 'Through the rapid acceleration of science and technology, man has acquired the power to transform his environment in countless ways and on an unprecedented scale'. If used 'wisely', the Declaration goes on to say, this power 'can bring to all peoples the benefits of development and the opportunity to enhance the quality of life'. If, however, it is applied 'wrongly or heedlessly', [it] can do incalculable harm to human beings and the human environment'. Apparently, human beings, while still being 'man', had done precisely what they were not meant to do, and the Declaration proceeds to list evidence of the 'unwise' and 'heedless' application of their power: 'dangerous levels of pollution in water, air, earth and living beings; major and undesirable disturbances to the ecological balance of the biosphere; destruction and depletion of irreplaceable resources' – the 'litany', as Lomborg (2001: 3) would years later call the long and often repeated list of environmental disasters, goes on.

In paragraph 4, the Declaration turns to the critical links between development and the state of the environment. 'In the developing countries most of the environmental problems are caused by under-development. ... Therefore, the developing countries must direct their efforts to development'. They must do so 'bearing in mind their priorities' – things such as 'adequate food and clothing, shelter and education, health and sanitation'. They must do so also bearing in mind what has become a priority to the countries for which adequate food and clothing, shelter and education, health and sanitation are not a priority, namely, 'the need to safeguard and improve the environment'. The 'developing countries', then, must develop because they lack what is essential for any society. They must develop also because their 'underdevelopment' causes environmental problems. But they must do so within reason. They should take measures to improve the general welfare of their population – that is their 'priority' – but must forget the slogans of the 1960s. Rather than a 'leap across the centuries', they should take only small and essential steps.

In paragraph 5, the Declaration returns to the links between 'underdevelopment' and the state of the environment and makes them more explicit. 'The natural growth of the population continuously presents problems for the preservation of the environment'. It does because more people means greater pressure on available resources. Hence, 'adequate policies and measures should be adopted ... to face these problems'. The Declaration does not elaborate further on the policies and measures to be adopted but the important thing to note here, in any case, is that the 'underdeveloped' are once again effectively told to forget the slogans of the 1960s with which the 'developed' tried to encourage and no doubt impress them. Certainly, gone is the idea that, as we have seen, the Earth could support a population as many as ten times more than the current level, and that the question was not 'how' – for there was no doubt that it could be done – but rather 'when'. In the emerging climate, such extravagant ideas had no place. Although seriously entertained only a decade earlier, they now seemed to belong to a very distant past.

Yet the most ironic reversal of all comes in the next paragraph, number 6. This is the point where the Declaration purports to clarify the meaning of freedom and to explain how it can now be achieved. 'For the purpose of attaining freedom in the world of nature', the Declaration says, employing a curious tone and lexicon strongly reminiscent of the earlier, modernist notion of 'the state of nature' – the argument, that is, that in such a state life can be nothing other than 'nasty, brutish and short', as Malthus famously put it – using such a tone and vocabulary, then, the Declaration has something quite different to say nonetheless: 'For the purpose of attaining freedom in the *world of nature*, man must use knowledge to build, in *collaboration* with nature, a better environment. To *defend* and improve the human environment ... has become an *imperative* goal for mankind' (UN 1973:

3; my emphases). We cannot know for certain why the authors of the Declaration chose to employ the strange locution 'the world of nature', but it does not seem to be simply a disinterested neologism. Given the drift of the argument, we may surmise that it was meant to correct, by reversing, the modernist assumptions about 'the state of nature'. Freedom will be achieved not by mastering nature – for it is no longer a 'state' and a predicament but a 'world' in its own right – but rather by collaborating with it. Such was now the definitive understanding of freedom, nature and the relationship between them. This was the truth recognised by everyone, the imperative goal for everyone. As it turned out, it was not recognised by everyone or, at any rate, not in quite the same way.

In 1973, in the aftermath of the Stockholm Conference on the 'Human Environment', and having by now received several puzzling and contradictory messages, the 'developing' countries felt the need to make a statement. By this time, the First Development Decade, which, as we have seen, was meant to carry out two revolutions at once, had come and gone without any revolutions taking place, while estimates for the Second Development Decade of the 1970s were 'extremely pessimistic' (Jankowitsch and Sauvant 1978: 217). In the Economic Declaration of the Fourth Conference of Heads of State or Government of Non-aligned Countries in Algiers, there is for the first time a section devoted to the environment. The first paragraph reads: 'The Heads of State or Government reaffirm their concern to ensure that the extra cost of environmental programmes should not prevent the fulfilment of basic development requirements, and they regard economic backwardness as the worst form of pollution'. And the second paragraph admits: 'They recognize that developing countries have their own environment problems which differ from those of developed countries and which require the attention of the international community' (Jankowitsch and Sauvant 1978: 224). Let us first note that here too the rhetoric of development is significantly scaled down. For the Non-aligned as much for the 'developed', development is no longer a question of 'leaps' and bounds but fulfilment of 'basic' requirements. Let us also note that the Non-aligned statement on pollution is not a recapitulation of the statement in the Stockholm Declaration about the 'Human Environment'. The latter was about economic backwardness as a cause of environmental pollution, the various pressures exerted on the environment through poverty, itself exacerbated by the increase in population. The Non-aligned statement, by contrast, is about economic backwardness as a *form* of pollution – indeed, as the statement suggests, a type of pollution worse than the pollution of the environment. What sort of pollution, then, did the Heads of State or Government of the Non-aligned have in mind? The statement cannot be understood outside the political and cultural context in which the 'developing' countries found, that is, both discovered and established themselves: the national and no doubt nationalist will to develop, modernise and attain as independent nations, if not the sort of 'pre-eminence' that Nehru dreamt for India, at least a certain standing in the community of nations. This wider context, then, may suggest possible answers to the question of poverty as a form of pollution.

To be sure, for all concerned, economic backwardness was responsible for poverty and disease. Hence, the priorities set in the Stockholm Declaration: adequate food, clothing, shelter, sanitation and health. But poverty and disease were themselves indicative of a certain attitude towards the world – an attitude that the 'developing' countries themselves or, at any rate, the élites in these countries, had come to denounce. As we have seen, both nineteenth-century modernisers and twentieth-century postcolonial élites presented this attitude as it was refracted through the modernist lens – as part and parcel of the moral economy of divine intervention, as an instance of how the metaphysical impinges on the physical and the social and hence as a reflection of the ignorance and superstition, or more kindly perhaps, the innocence of local populations. People accepted their predicament, the rhetoric went, because they perceived it as the will of God. It would seem, then, that economic backwardness was worse than environmental pollution because it was indicative of 'men' at their worst – still traditional, still backward, still credulous and superstitious. It was the worst form of pollution because it was indicative of a condition not worthy of what the 'developing' aspired to become – modern 'men' that belonged to modern nations. It was the worst form, in short, because it was indicative of cultural pollution. If so, it was imperative for the 'developing' countries to make a statement. It was imperative to defend their will to develop and modernise even if, by doing so, they were also defending, no doubt unwittingly, the Western cultural assumptions that rendered economic backwardness the worst form of pollution and hence the conditions also responsible for their inferior position in the global hierarchy. It was imperative to make their views known to the 'developed' countries, even when, as it was already beginning to happen, a new item appeared on the list of indicators of cultural pollution which threatened to expose them in the eyes of the 'developed' yet again, namely, that 'man' is at his worst *also* when he pollutes the environment.

Yet these were still early days in the struggle for modernisation, development and national standing. As the new ontology consolidated itself, the 'developing' began to come to terms with the new item on the list of cultural pollutants and to employ it strategically to their advantage. It is within this wider context of struggle that the second part of the statement by the Heads of State or Government of the Non-aligned should be understood. There is no doubt that their recognition of the existence of environmental problems in both 'developed' and 'developing' countries was partly the result of logical conformity to a set of cultural assumptions: that there are 'facts' about the world which science is capable of discovering and delivering in an objective fashion; that environmental degradation as a scientific 'fact' cannot be disputed, even if the extent of it perhaps could; and that insofar as the problems besetting the environment have been discovered by Western scientists, this was understandable and to be expected. Western countries were far superior in scientific matters than the rest of the world – not in matters of the 'spirit' but certainly in matters pertaining to the 'external', material domain. Beyond logical conformity, however, there is the wider cultural and political context to be taken into consideration. As I will argue in detail in the last chapter, the meaning that the 'developing' countries attach to the 'environmental crisis' is not quite the same as that attached by the 'developed'. At the level of national governments and bureaucracies, it is shaped by the strategic deployment of the 'crisis' as means of undermining Western hegemony and, by the same token, a means also of buttressing the non-West's presumed superiority in matters of the 'spirit' – a materialist West has damaged nature almost beyond repair, a spiritual East will provide guidance as how best to heal it. It is also shaped by the need to pursue economic development. The 'environmental crisis' makes sense to the 'developing' because their 'underdevelopment' contributes to it and underscores the urgency for development.

The important thing to note here, in any case, is the balancing act that the 'developing' countries were from now on forced to carry out in their struggle for development and national prestige. The problem to be negotiated was a set of conflicting values: first, that poverty is cultural pollution because it is indicative of superstition – which is the ignorance of backwardness – and second, that environmental pollution is cultural poverty because it is indicative of the refusal to come to terms with the 'truth' about the nature of nature – which is the ignorance (and arrogance) of 'man'. In successfully negotiating this hegemonic double bind, the 'developing' countries also reproduced successfully all the cultural conditions responsible for it and secured their further and deeper entanglement in it. This balancing act was publicly recognised fourteen years after the Non-aligned Conference in Algiers and was announced in *Our Common Future*, the Report of the World Commission on Environment and Development (WCED), commissioned by the United Nations and presented to the General Assembly in 1987. It was formalised and institutionalised as 'sustainable development', which the Report defines vaguely as 'development that meets the needs of the present without compromising the ability of future generations to meet their own needs' (WCED 1987: 43). And it received the stamp of approval of the 'developed' who had their own balancing act to perform: how to negotiate successfully between the imperatives of geo-political power – the need to retain their dominant position in the community of nations – and the imperatives of the environment. With the institutionalisation of this double balancing act as 'sustainable development' the obstacles were removed and the stage set for another round of the power game in which the West and its Others have been engaged for the last several centuries. The complicity that united them in disagreement all this time was silently but firmly renewed.

I shall examine the complicity of the 'developing' at the level of national government in more detail in the last chapter. Enough has been said here also to highlight the extent of the change that is currently upon us – the transformation of 'nature' into nature and 'man' into the 'human being' – as this is reflected in the 'mainstream' version of the environmentalist vision. A fuller picture of this momentous change cannot be attained without the radical environmentalist view and without taking into account the change in the perception of those people who were until recently regarded as 'primitive' and 'backward'. What must also be included in the picture is the perception of the critics of environmentalism, the defensive and, as we shall see, misinformed position of the apologists of the modernist paradigm. I shall turn to the first two below and to the third in the next chapter. For the moment, it may be pertinent to raise a few preliminary questions regarding the condition of possibility of the environmentalist vision of the world. I shall do so, in the first instance, by going back to the beginning of the Report of the World Commission on Environment and Development and to an often-quoted passage that develops further the theme of 'fragile' nature through the use of a wider perspective.

> In the middle of the 20th century, we saw our planet from space for the first time. Historians may eventually find that this vision had a greater impact on thought than did the Copernican revolution of the 16th century … From space, we see a small and fragile ball dominated not by human activity and edifice but by a pattern of clouds, oceans, greenery, and soils. Humanity's inability to fit its doings into that pattern is changing planetary systems, fundamentally. Many such changes are accompanied by life-threatening hazards. This new reality, from which there is no escape, must be recognized – and managed. (WCED 1987: 1)

On first reading, the reading intended by the authors of the Report, this passage says something along the following lines: Seeing our planet from space was an event of the highest historical and philosophical significance. For the first time, we were able to see the Earth for what it really is: 'a small and fragile ball'. Having seen it in its true light, we realise with greater urgency than ever before that we must fit our doings within the planetary systems. We have caused them enough damage as it is, and this is already threatening life on the planet. This is the 'new reality' and there is 'no escape' from it. It must be recognised by everyone and managed accordingly. Perhaps there is no escape from this 'new reality'. But if so, it behoves us to examine it in detail and question its claims with all due seriousness. If we cannot escape it, we should at least know what it is exactly we are destined to live with. Let us, then, attempt a second and more discriminating reading of this passage.

The perspective from space produces a field of total visibility, which is also, at the same time, a field of total invisibility. Looking from such a distance, two things happen at once. On the one hand, we can see the whole of the planet, on the other, we lose sight of the human being. What we see is not human 'activity' and 'edifice' but a pattern of clouds, oceans, soils and so on. This is to be expected, of course. Compared to the planetary systems – the lithosphere, hydrosphere, biosphere and atmosphere – human beings and their products are far too small to be visible. And yet, although they are too small and the planet too large compared to them, the Report also claims that human beings are too big and the planet too small for them. Although it is not human edifice and activity that 'dominate' the planet, it is in fact human edifice and activity that dominate it – as the Report insists, human activity is changing the planetary systems 'fundamentally'. And although what we see from space is the enormity of the planetary systems – which drives home the message that we must fit our 'doings' in their 'doings' and not the other way round – we also see that the planet is 'a small and fragile ball'. On this more discriminating reading, then, it is not at all clear what it is that we see from space. If anything is clear, it is that the Report's claim about what we see is riddled with several contradictions – which are the result, as we shall subsequently see, of mixing the literal with the metaphorical.

The second issue to raise here has to do with scientific facts, those marshalled to support the vision of the 'fragile planet'. I have already suggested that the argument of facts is hardly adequate to explain the environmentalist perception of the world. To elaborate further: 'In the early 1970s', writes Lovelock in the Preface of *Gaia*, 'we were still innocent about the environment'. To plead innocence is, of course, to plead ignorance, but what sort of ignorance might this be? 'Rachel Carson had given us cause to worry', Lovelock goes on to say; 'farmers were destroying the pleasant countryside we knew by the overuse of chemicals but it all still seemed alright. Global change, biodiversity, the ozone layer, and acid rain all were *ideas barely visible* in science itself, still less of public *concern*' (2000: vii; my emphases). Such things were 'barely visible' in the 1970s, then, but is this to say they were virtually invisible because they were just beginning to emerge or because, although they had already emerged, there were no categories of perception to render them visible? Lovelock does not raise this question but it is interesting to note that he refers to global climatic change, the ozone layer and acid rain as 'ideas' rather than, as one might expect, concrete, empirical phenomena. A slip of the tongue, one might say. Perhaps. Yet, even if unwittingly, Lovelock is not far off the mark. Not that acid rain, the greenhouse effect or the depletion of the ozone layer can be dismissed as fictions of the environmentalist imagination. He is not far off the mark, rather, because realities emerge and become visible, relevant and meaningful, within determinate cultural contexts, because 'facts' are noticed by those who are predisposed to take notice and become a cause of concern to those who are already in a state of concern. That environmental danger is a matter of cultural perception has been pointed out often enough – most famously by Douglas and Wildavsky (1982) and Douglas (1994) – and there is really no need to labour the theoretical point further, except perhaps by giving it a more concrete, empirical form. Let us, then, raise the question that environmentalists themselves often raise, albeit in the context of what they perceive as the apathy of the general public: if scientific 'facts' are so transparent, if they are self-evident truths and speak by themselves, why do people hear them differently or, indeed, why do some people not hear them at all? Why is it that their facticity does not impress itself upon everyone in the same way?

The usual explanation, of course, is that people are ignorant of the extent of the 'environmental crisis', in which case they must be educated, or that they are too narrow-minded and too selfish to be concerned with anything other than their personal interests, in which case they may have to be forced to comply. Yet as many environmentalists also recognise, this may not necessarily be the case. 'Opinion polls confirm that many people now accept the significance of the crisis we call environmental ... Yet individually, relatively few people do anything significant about this recognition in their personal lives' (Grove-White 1993: 29; see also Callicott 1999). Although many people are aware of environmental problems and recognise their importance, then, most do not seem prepared to engage them seriously. Such is the 'troubling paradox', says Grove-White. 'Can it be that such inconsistency reflects in part the inauthenticity of the largely "physicalist" descriptions of what is at stake?' Are the facts about the 'environmental crisis', in other words, so 'factual' and 'scientistic' as to fail to move the wider public? Is there something else 'at stake' over and above the environment, something more 'authentic' than scientific facts that may tip the balance? Grove-White thinks so. 'Scientism', he points out, has failed to 'engage people's full being'. It is inadequate, he says in another section of his paper, because of its 'superficial treatment of the mysteriousness and open-endedness of existence itself. There is little sign in the official descriptions of environmental problems or methodologies of the radically unknown character of the future, or of humankind's place in creation' (Grove-White 1993: 23–24). Although the 'mysteriousness and open-endedness of existence', the 'radically unknown character of the future' and 'humankind's place in creation' are seemingly unrelated to environment concerns, it is such things, according to Grove-White, that make all the difference and account for the difference between mere knowledge and conviction, indifference and concern, apathy and action.

Science, then, appears unable to relate the whole story about environmentalist perception. If what Grove-White says is anything to go by, it can tell us very little. It can only point to facts, but facts by themselves are not enough to explain effective engagement with the world. As the prominent Norwegian environmental philosopher Arne Naess (1989: 24) put it, 'chemistry, physics, and the science of ecology acknowledge only change, not valued change. ... We need another approach which involves the inescapable role of announcing values, not only "facts"'. What is needed over and above science is something that captivates people's 'full being' – a system of values, a moral story, an ontological master-narrative within which the 'environmental crisis' becomes not only visible, but also relevant and meaningful – a uniquely realistic proposition. But perhaps Grove-White and Naess are overstating their case against the scientism and ontological poverty of environmental discourses. As Szerszynski (1996: 105–106) argues, 'even the dominant modernist versions of environmentalism' – the versions, that is, which rely heavily on science to make their case – 'and even when they take their most technocratic and seemingly morally neutral guises, can themselves be seen as crucially relying on certain moves within [the modernist] problematic' – the problematic, that is, in which people strive to 'find meaning in a meaningless universe'. There is plenty of evidence to support Szerszynski's claim. Here, I shall look at some examples from the United Nations literature.

Let us first return briefly to the WCED Report, *Our Common Future*. 'The planet's species are under stress', says the Report in one of its many warnings, and then goes on to explain why biodiversity is important. 'The genetic material in wild species contributes billions of dollars yearly to the world economy in the form of improved crop species, new drugs and medicines, and raw materials for industry'. Yet although important, dollars are not the only issue here. 'Utility aside, there are also moral, ethical, cultural, aesthetic, and purely scientific reasons for conserving wild beings' (WCED 1987: 13). In the chapter on Species and Ecosystems, the Report returns to the same theme. 'Species conservation is not only justified in economic terms. Aesthetic, ethical, cultural, and scientific considerations provide ample grounds for conservation'. Aesthetic, ethical and cultural considerations are not quantifiable, however. For the benefit of those who are more comfortable with facts and figures and 'demand an accounting', therefore, the Report returns to purely economic considerations to reiterate that 'the economic values inherent in the genetic materials of species are alone enough to justify species preservation' (WCED 1987: 155). The Report does not elaborate on the nature of the 'moral, ethical and cultural' considerations, no doubt, partly because many people still 'demand an accounting'. Yet in another, more recent publication sponsored by the United Nations, we acquire a better sense of what might be involved in these non-quantifiable considerations.

Caring for the Earth: A Strategy for Sustainable Living is a joint publication of The World Conservation Union (IUCN), the United Nations Environment Programme and the World Wide Fund for Nature, which sets out the principles of sustainable living. The first principle – 'the founding principle providing the ethical base for the others' – is 'respect and care for the community of life'. As the Strategy says, 'this principle reflects the duty of care for other people and other forms of life, now and in the future. It is an ethical principle' (IUCN *et al.* 1991: 9). Respect and care, then, not just for human life but for other forms of life as well. This ethical principle is itself based on the recognition that 'nature has to be cared for in its own right, and not just as a means of satisfying human needs'. Beyond the value that it has for human beings, nature as a domain of life has value in and of itself. It is this value, which is independent of human beings, that makes care for nature 'morally right' (IUCN *et al.* 1991: 13). The belief that 'every life form warrants respect independently of its worth to people' (IUCN *et al.* 1991: 14), which is an extension of modernist ethics, is known in the literature of radical environmentalism as 'biocentrism'. As we shall see, it is a cardinal value, and the value that radical environmentalists often point to in order to differentiate and distance themselves from 'mainstream' environmentalism. If the foregoing assertions are anything to go by, they may be exaggerating the difference.

Since ethics and science are not known to mix very well, the authors of *Caring for the Earth* turn for support to that other institution which, if Durkheim is to be believed, although progressively displaced by science has something 'eternal' about it nonetheless. 'Establishment of the ethic needs the support of the world's religions because they have spoken for centuries about the individual's duty of care for fellow humans and of reverence for divine creation' (IUCN *et al.* 1991: 13). The world religions were only too happy to oblige.[3]

'This Sacred Earth'[4]

Differences aside, some of which are real, some imagined and some simply rhetorical, radical environmental discourses and practices – 'radical ecologies' for short – are united by the same fundamental aim: 'they strive to liberate all life' (Zimmerman 1994: 318).[5] They are united also by the only thing that could possibly give rise to such an aim – an ethics and a metaphysics that imbue the ecology of the 'radical ecologies' with certain spirituality and transform the fragile nature of 'mainstream' conservationism into consecrated ground, 'Planet Earth' into 'Sacred Earth'.

Differences aside, the radical ecologies are also united by their distinctive critique of 'man'. If in the discourses produced by the United Nations experts, 'man' could get away with a reproach, however severe, for being improvident and unwise in squandering the planet's scarce resources, in the radical ecologies he is up against something far more serious and damning. Human beings are neither stewards of the planet nor in any ontological sense different from nature. This is a grave misconception, according to radical environmentalists, and the mark of profound alienation. Human beings are denizens of the planet and part of nature, indeed, a very small part of a wider, ongoing, possibly providential, certainly grand and extraordinary process. They are part of the process by which Life unfolds, reproduces and propagates itself – a process that has its own logic and whose ultimate meaning human beings do not comprehend and should not interfere with. Human beings are but a drop in this vast ocean of Life, one creature of Creation among countless others, whether 'Creation' is understood in Christian terms, as in 'eco-theology' (e.g. Berry 1995), or in terms of evolutionary biology and cosmological theory. To have thought otherwise is the mark not of improvidence but hubris, of utter lack of respect towards something immensely larger and grander than human beings. 'It is *hubris* to declare that humans are the central figures of life on Earth and that we are in control. In the long run, *Nature is in control*' (Spretnak 1984: 234). It is hubris and bears witness to the outrageous arrogance of 'man' – 'the arrogance of humanism' (Ehrenfeld 1978).

It is not my intention in this chapter, or anywhere else in the book for that matter to ridicule radical ecologies as a mythical or mystical discourse. If it appears so, this is only because one has to use a language – the only one available – that is already predisposed to dismiss the mythical and the spiritual. Even anthropology, the most sympathetic of all social sciences to such systems and to those who practise them, can do nothing better than to render them in more 'rational' terms and hence, in effect, to explain them away. Certainly, my aim is not to contrast the radical ecologies with conservationist approaches and other scientific ecologies in order to defend science and reason against unreason. As I have already suggested, scientific ecologies cannot always hide their ethical underpinnings, while radical ecologies themselves straddle both science and non-science.[6] At any rate, as we shall see in the next chapter, the defence of reason against the presumed unreason of the radical ecologies has been undertaken by others for whom these terms are self-explanatory and who have stakes invested in their strict separation. My aim is quite different and rather more complex. As a first step, it is to provide a phenomenological description of the extent of change in the perception of nature and humanity since the early 1950s. But in a second and related move, it is also to go beyond phenomenology and scrutinise the nature of this transformation. As I will argue in subsequent chapters, this change does not constitute a shift from reason to unreason, nor is it partly or even mainly a shift from science to ethics and metaphysics. Nor does it reflect a reversal in the order of things, as a phenomenological description may suggest. It is, rather, a change in the same direction, an intensification, greater objectification and totalisation of the *same* order – an order re-ordered and reorganised by having been taken to its logical and onto-logical extreme. In a social universe whose cultural logic is to strive constantly for ultimate universalisms, this change stands witness to the fact that the last grand division of the Whole – the division between humanity and nature – has finally been brought 'into the focus of European thought' and serious efforts are being made to efface it. This is to say also that although this change does constitute a difference, it is a difference that in a fundamental sense – both in terms of the nature of the modernist paradigm itself and the nature of the power relationship between the West and its Others – makes little or no difference.

It should not come as a surprise, then, that the 'Life' which radical environmentalists strive to liberate is the whole of life, life writ large, indeed, more than what is ordinarily understood by 'life'. Unlike the luminaries of the eighteenth and nineteenth centuries whose primary concern was humanity and who, given the grand division between the human and the non-human, conceptualised liberation in terms of setting 'man' free from the 'shackles' of internal and external 'nature', radical environmentalists are concerned with the liberation of everything. They wish to liberate human beings from their 'misconceptions' about the true nature of Being – the 'truth' that all Life is One – and their alienation from themselves, other human beings and the non-human world itself. The 'Life' with which radical ecologies are concerned includes everything that the term signifies in everyday discourse, from the smallest and seemingly most simple and insignificant plant and animal to the largest; it includes everything that the more technical biological definition of life designates; and it goes beyond both to encompass everything that supports life – rivers, lakes, oceans, mountains, the soil and the air. In the technical language of ecology, 'Life', as radical environmentalists understand it, refers to the ecosphere rather than merely the biosphere. In the equally technical but more colourful language of the radical ecologies, it refers to 'Gaia', the Earth Goddess or 'Mother Earth', one vast and vastly complex organism, the living, self-generating and regenerating planet.

As Arne Naess (1989: 29) points out, the term 'Life' is used 'to refer also to things biologists may classify as non-living: rivers (watersheds), landscapes, cultures, ecosystems, "the living earth". Slogans such as "let the river live" illustrate this broader usage so common in many cultures'. A river condemned to death is a river whose natural flow is arrested by human intervention, trapped in a dam, prevented from reaching its ultimate destination and hence prevented also from fulfilling its destiny or *telos* – a river, in short, which is not allowed to be itself. A river is a good metaphor to use in trying to explicate what radical environmentalists have in mind when they set as their goal the liberation of all life. Like a river, life flows. It unfolds itself and follows its own course – its evolutionary destiny – according to a logic that is independent of what human beings may think or would like. There may or may not be an ultimate meaning and purpose in the evolution of life on Earth. The point, according to radical environmentalists, is that each life form endeavours to preserve and propagate itself and this is good enough a reason to grant it respect. This becomes even more compelling when one realises that life is not passive existence but creative striving. Naess articulates this idea of active involvement of life forms in their own (well) being with the notion of 'Self-realisation'. The conventional idea of self-preservation, he points out, 'is misleading in so far as it does not account for the expansion and modification. There is a tendency [in all life forms] to realise *every* possibility for development'. Naess traces the notion of 'Self-realisation' to

Spinoza's conception of *perseverare in suo esse*, 'to persevere in one's own (way of) being, not mere keeping alive' (1989: 166); to realise one's full potential and flourish, not mere surviving. Murray Bookchin, the founder and main spokesperson for social ecology, argues along similar lines. 'Life is not passive in the face of ... possibilities for its evolution. It drives toward them actively in a shared process of mutual stimulation between organisms and their environment ... as surely as it also actively creates and colonizes the niches that cradle a vast diversity of life-forms in our richly elaborated biosphere'. The fact of striving 'imparts an identity, indeed, a rudimentary "self", to every organism'. It means choice, and choice means 'the rudimentary fact of *freedom*' (Bookchin 1993: 105). What right, then, do human beings have to stem the flow of life? How can they deny other life forms the freedom to be themselves?

As radical environmentalists argue, the idea that life is a creative striving is supported by recent rethinking in systems theory and evolutionary biology. Capra (1996: 217), for instance, points out that 'we are beginning to recognize the creative unfolding of life in forms of ever-increasing diversity and complexity as an inherent characteristic of all living systems'. Mutation and natural selection, the cornerstones of Darwinian evolutionary theory, are still considered important but with the recent rethinking 'the central focus is on creativity, on life's constant reaching out into novelty'. What is completely gone is the old idea of the 'survival of the fittest'. 'There is no such thing as the "fittest" kind of organism', according to one of the leading American cell biologists. 'We can only talk about how an organism propagates in a given niche. ... It is no more or less fit than another kind of organism that has adapted to some other niche' (Goodenough 1998: 78). An appropriate metaphor for this non-hierarchical evolutionary theory is the branching tree. As Fox (1990: 225) points out, 'we can think of ourselves and all other presently existing entities as leaves on this tree', neither better nor worse than other leaves, all embodying the same value and worth. What evolution produces, then, is not superior and inferior life-forms but a biosphere of an ever-increasing complexity and diversity.

The radical environmentalist effort to liberate all life is grounded in both ethics and metaphysics. It is an ethical undertaking with metaphysical underpinnings even when, as in the case of ecofeminism, an attempt is made to be ethical without employing a wider, cosmological, and according to ecofeminists 'androcentric' and patriarchal perspective. In the deep ecology of Arne Naess and his followers, all beings, whether living or non-living in the strict biological sense, have intrinsic value, which is to say, value independent of human evaluation. And because they have value in and of themselves, all beings are morally considerable and deserve respect for what they are. Naess and George Sessions formalised this idea in a 'platform' for radical environmentalism. The first point of the platform reads: 'The well-being and flourishing of human and non-human life on Earth have value in themselves (synonyms: intrinsic value, inherent worth). These values are independent of the usefulness of the non-human world for human purposes' (Naess 1995a: 68). The idea of intrinsic value in nature is a matter of considerable debate among radical environmentalists. Some environmental philosophers, such as Callicott (1999), Zimmerman (1994) and Fox (1990), argue that there can be no such thing as objective value in nature, no value independent of human intention and evaluation. What this means is that to grant moral consideration to non-human beings on the basis of (inevitably human) value is to fall into the trap of 'anthropocentrism'. It is to make the non-human world dependent for its being on the 'human being'. For these philosophers respect for nature is not a matter of rights, nor can it be a moral imperative. It is an ontological question – the question of comprehending the nature of Being, which, as we have seen, is that all life is essentially and fundamentally One. 'Ontology precedes ethics', according to Zimmerman (1994: 109) – knowing what is, is prior to knowing how one ought to behave. If ontology precedes ethics, respect for nature becomes spontaneous and natural.

Although Naess is not prepared to abandon completely the idea of intrinsic value in nature, he has no real means of defending it. 'The burden of proof', he points out, 'lies with the subjectivists' (1993: 35). At any rate, for Naess, as much as for other radical environmental philosophers, respect for nature cannot be a moral injunction. The aim of the radical ecologies is not to prove objectively the existence of intrinsic value, in which case people would presumably feel obligated to be morally considerate to non-human beings. Rather, it is to encourage 'identification' with nature, since belief in intrinsic value is itself the outcome of this process. Being a subjective experience, identification falls outside the domain of strict rationalism and empirical proof. 'It is a spontaneous, non-rational, but not irrational, process', a process through which the interests of another being become one's own (Naess 1993: 29). Having identified with another being, one can have no doubt about its intrinsic value and worth. Having had such an experience, one would care for the other being not because of some onerous moral obligation but naturally and spontaneously. To identify with another being, whether human or non-human, is to see oneself in that being. It is to experience intense empathy for the other, joy in its well-being and development, admiration for its doings, sympathy and compassion for its suffering. For Naess (1993: 32), suffering 'is perhaps the most potent source of identification'; it is what makes human beings relate to other beings more readily. His paradigmatic example refers to the suffering and eventual death of a very small and seemingly insignificant being.

> My standard example involves a nonhuman being I met forty years ago. I was looking through an old-fashioned microscope at the dramatic meeting of two drops of different chemicals. At that moment, a flea jumped from a lemming which was strolling along the table and landed in the middle of the acid chemicals. To save it

was impossible. It took many minutes for the flea to die. Its movements were dreadfully expressive. Naturally, what I felt was a painful sense of compassion and empathy. But empathy was *not* basic, rather it was a process of identification: that 'I saw myself in the flea'. If I had been *alienated* from the flea, not seeing intuitively anything even resembling myself, the death struggle would have left me feeling indifferent. So there must be identification in order for there to be compassion and, among humans, solidarity. (Naess 1995b: 227)

Unlike many people, Naess could identify with the dying flea. Because he saw himself in the flea, he could not help but feel intense empathy and compassion. A case of extreme sensitivity, one might say. Perhaps. Yet the issue for radical environmentalists of Naess persuasion is not so much psychological as ontological. As I have already pointed out, the central premise of the radical ecologies is that the division between the human and the non-human is a false dichotomy. Most people cannot identify with a flea because they are trapped into a narrow and, radical environmentalists would say, egotistical self. They perceive of the self and the flea not only as two distinct entities, which they are, but also as two entities that have nothing in common. This is a sign of profound alienation. Life is fundamentally One, and because it is one and the same, one should experience 'oneself to be a genuine part of all life' (Naess 1989: 174). Achieving this sort of understanding about the nature of Being, which Naess links to Indian philosophy and particularly to Gandhi, constitutes Self-realisation for human beings. People reach their full potential only when they develop as wide a sense of the Self as possible. 'Self-realization in its absolute maximum ... is the mature experience of oneness in diversity' (Naess 1993: 28).

Naess's key notions of Self-realisation and identification have been both quite influential and controversial among radical environmental philosophers and activists. They have been embraced and further elaborated but also contested and criticised, most notably by ecofeminists. Warwick Fox, one of the leading supporters of Naess's views, distinguishes between three types of identification. The first and simplest kind develops through personal involvement with other beings, whether human or non-human. One identifies with what one knows personally. The other two kinds, 'ontological' and 'cosmological' identification, are 'transpersonal'. The expanded self is able to identify with beings with which it has no personal relationship or direct contact. In its maximum extension, the self is able to identify with everything that is. Fox links the first type of transpersonal identification to the ontology of Martin Heidegger. Ontological identification, Fox admits, 'is not a simple idea to communicate with words!'. It has to do with the seemingly obvious but in reality quite enigmatic 'fact *that* things are'. Out of the 'background of nonexistence, voidness, or emptiness', beings emerge into the world. 'Things *are*! There is something rather than nothing! Amazing!'. This 'utterly astonishing fact ... impresses itself upon some people' to such an extent as to generate a sense of 'the specialness or privileged nature of all that exists'. It is this sense of 'specialness', it would seem, the 'privilege' of being vis-à-vis the utter nothingness of non-being, which makes it possible for some people to identify with everything that is (in being). The sheer fact of existence is for some people the fundamental link that connects everything together. Fox tries hard to explicate a highly abstract, intuitive and no doubt metaphysical idea, not with a great deal of success. In the end, he resorts to a simpler course of action. 'I can only reiterate that these remarks cannot and should not be analyzed through a logical lens. We are here in the realm of ... the mystical' (Fox 1990: 250–51).

Michael Zimmerman, who favours ontological over cosmological identification as the basis for opening up the narrow egotistical self, provides a rather different interpretation.

> Ontological identification is made possible when the ego-subject is revealed not as a solid entity, but as a shifting and changing phenomenon. ... Often associated with this revelation is the insight that all spatiotemporal phenomena arise 'within' an all-encompassing, generative 'emptiness' (*sunyata*), sometimes called 'absolute nondual consciousness'. Mystics ... assert that 'enlightenment' involves identifying with this generative nothingness. ... By identifying with generative, absolute nothingness, it becomes simultaneously possible both to affirm and to show compassion for all phenomena that arise and pass away 'within' such nothingness. (1994: 53)

It would seem, then, that for Zimmerman what makes identification possible is not the 'privilege' of existence but its reverse, the fact of non-being. The 'enlightened' self identifies with the absolute void, the nothingness that generates everything. Having in this way become nothing itself, an unimaginably vast and empty space, this 'self' can accommodate everything. It becomes all phenomena that arise and pass away and all phenomena become it. No wonder, then, that the 'enlightened self' can affirm and show compassion for everything.

The last type of transpersonal identification, which is Fox's preference, has its basis in cosmology. In principle, this could be any kind of cosmology, according to Fox – mythical, religious, philosophical or scientific. Whatever its nature, what makes cosmology a sound basis for wider identification is the idea that everything which exists is ultimately traceable to the same origin. Depending on the cosmology, the origin of everything could be the union of the earth and the sky, as in the mythology of many native people, the God of the Judeo-Christian tradition, Spinoza's Nature/God, the Big Bang of cosmological theory, or the First Cell of evolutionary biology. Whatever it is, the same origin renders all beings essentially and fundamentally the same. For Fox, the preferred cosmology for the environmentalist movement is the cosmology of physics and biology. 'Modern science is providing an increasingly detailed account of the physical and biological evolution of the universe that compels us to view reality as a single unfolding process' (Fox 1990: 253). Biology teaches us that every living being originates from the same cell. On a much grander scale, modern physics teaches us that everything in the universe originates from the same mathematical point. This makes us human beings the same not only with everything that exists or has ever existed on our planet but also with everything that exists or has ever existed in the universe writ large. We are the same, not only with animals, trees, plants, rivers and mountains but also with stars and galaxies. Everything is made of the same cosmic particles, the very same star-stuff. 'All that is now', Seed and Fleming (1996: 503) point out meditating on the Big Bang, 'every galaxy, star and planet, every particle existing comes into being at this great fiery birthing. Every particle which makes up you and me comes into being at this instant and has been circulating through countless forms ever since'. For environmentalists of Fox's persuasion, this realisation is the highest form of ecological consciousness, the absolute maximum of Self-realisation – a self so wide that can encompass the entire universe. This universal, cosmic self is also the bone of contention between deep ecology and ecofeminism.

The ecofeminist critique of the self that identifies with everything is rooted in gender politics. Ecofeminists view the destruction of nature and the subordination of women as two inextricably interrelated phenomena. The link is none other than man himself – 'man' the master of 'nature' and man the patriarch and master of women. Historically, the emergence of science and technology spelled the end of the organic model in which nature was perceived as a nurturing mother and replaced it with a mechanistic model in which nature (still female) was perceived as dead matter. This shift – the 'death of nature' as Merchant (1980) would have it – helped to legitimise not only the opening up of nature for exploitation but also the perpetuation of the sociocultural conditions responsible for the subordination of women. Synchronically, patriarchy as a conceptual framework and a social order operates on the basis of a 'logic of domination' in which things traditionally associated with men are accorded higher status and prestige than things traditionally associated with women. This logic 'puts men "up" and women "down", culture "up" and nature "down", minds "up" and bodies "down"' (Warren 1987: 6). It valorises reason at the expense of emotion, the abstract over the concrete, the universal over the particular, the cognitive over the experiential, in short, men over women. In some versions of ecofeminism women are closer to nature by nature (Salleh 1984); in others, female attributes are the result of culture and socialisation. In both, it is only the female way of being – which is to say, being attuned to emotions and the body, mindful of experience and attentive to the particular and the concrete – that offers any real hope of overcoming the current 'environmental crisis'.

Given this perspective, it should come as no surprise that ecofeminism is highly critical of everything in environmental discourse that appears rationalistic, abstract, disembodied and universalising. Under attack, for example, are all axiological theories that posit intrinsic value in nature and then universalise an ethic of respect based on this theoretical assumption. Axiological approaches are 'cognitivist', according to ecofeminists; they treat ethical behaviour as the outcome of rational understanding. As such, they are also sexist. They reproduce the familiar dichotomy between reason and emotion by treating '"desire", caring, and love as merely "personal" and "particular" … "feminine" emotions as essentially unreliable, untrustworthy, and morally irrelevant, an inferior domain to be dominated by the superior, disinterested (and of course masculine) reason' (Plumwood 1998: 293). Under attack also, for similar reasons, is deep ecology, particularly the idea of transpersonal identification. As Spretnak (1997: 428) points out, the notion of an expanded self that can identify with everything 'is immensely unappealing [because] it can be construed to mean the expansion of the male ego to cosmic proportions'. Ecofeminists propose an ethic of respect for nature based not on the expanded self but on the 'self-in-relationship'. The former obliterates the distinctiveness and independence of other beings by incorporating everything into the enlarged but egotistical (male) self. The latter respects difference by maintaining both a sense of self and a sense of the other beings' distinct identity. The former produces 'a vague, bloodless, and abstract cosmological concern' for nature (Plumwood 1998: 303); the latter generates deeply-felt emotions – of care, sympathy, trust, friendship, love – for non-human beings which are the outcome of personal encounters and involvement. Rather than impersonal identifications, then, ecofeminists call for personal relations – 'care and responsibility for particular animals, trees and rivers' – as the only sound basis for acquiring a wider concern and respect for nature (Plumwood 1998: 295).

These differences between ecofeminism and other radical ecologies aside, the ethics and the metaphysics that underpin both produce scenarios of encounters with nature which from both a rationalistic and common sense perspective – the perspective, that is, of a culture still accustomed to treating nature as a disenchanted domain – appear alien, mystical, irrational and hence also to a certain extent disturbing and threatening. Although there are countless examples of such scenarios, it is important to explore at least some of them. I should perhaps reiterate that my aim here is not to confirm the claims of reason and common sense. Such would be the position of those who believe that what is at stake in the struggle between the environmentalist and modernist paradigms is 'the truth' about the nature of reality. I hold a different view. What is at stake is 'a truth', a historically and culturally contingent field of relevance and meaning whose stakes are meaning, in the ontological sense, identity and power. My aim, then, is simply to highlight the extent of change in the perception of nature and humanity over the last few decades. And this as a preamble to the more difficult and more fundamental question regarding what lies behind this phenomenon.

I shall begin my exploration of encounters between radical environmentalists and non-human beings with an example of personal identification, the kind of identification propagated by ecofeminists, which involves a climber and a rock.

> On my second day of climbing … I looked all around me – really looked – and listened. I heard a cacophony of voices. … I closed my eyes and began to feel the rock with my hands. … At that moment, I was bathed in serenity. I began to talk to the rock in an almost inaudible, child-like way, as if the rock were my friend. I felt an overwhelming sense of gratitude for what it offered me – a chance to know myself and the rock differently … to know a sense of *being in relationship* with the natural environment. I felt as if the rock and I were silent conversational partners in a longstanding friendship. I realised then that I had come to care about this cliff. … I wanted to be with the rock as I climbed. Gone was the determination to conquer the rock. … I wanted simply to work respectfully with the rock as I climbed. (Warren 1998: 332)

Here, then, is a revelation, the unexpected discovery of a hidden truth, a truth as much about the rock as about the climber. The rock is another being, different from ourselves, no doubt, but a being that we can identify with nonetheless. In our haste to climb and triumph over the rock, to conquer it by reaching the summit, we have forgotten that it can be a conversational partner, albeit a silent one, a friend to work and be with, something to care about and be thankful to. We have forgotten that we can connect with the rock, be in harmony with the natural environment and in touch with our own true selves. Such is the extent of our alienation.

The second example is not quite an encounter, but it does nonetheless illustrate quite graphically the wider, cosmological and ontological identification with everything that is. It is an invitation to an encounter or rather an 'invocation'. As for the being in question, this is not anything tangible – a rock, a tree, an animal. The supplicants ask for the presence of the 'spirit of Gaia'.

> *We ask for the presence of the spirit of Gaia* and pray that the breath of life continues to caress this planet home.
>
> May we grow into true understanding – a deep understanding that inspires us to protect the tree on which we bloom, and the water, soil and atmosphere without which we have no existence.
>
> *We ask for the spirit of Gaia* to be with us here. To reveal to us all that we need to see, for our own highest good and for the highest good of all.
>
> We call upon the spirit of evolution, the miraculous force that inspires rocks and dust to weave themselves into biology. You have stood by us for millions and billions of years – do not forsake us now. …
>
> Awaken in us a sense of who we truly are: tiny ephemeral blossoms of the Tree of Life. Make the purposes and destiny of that tree our own purpose and destiny.
>
> Fill us with love for our true Self, which includes all the creatures and plants and landscapes of the world. Fill us with a powerful urge for the well-being and continual unfolding of *this* Self. (Seed 1996: 499)

To establish a relationship with the rock, to invoke the spirit of the Earth and pray to it for enlightenment and guidance: in the history of anthropology views of the natural world such as these were for a long time held to be indicative of the 'irrationality' and 'ignorance' or, in a more patronising vein, the innocence of 'primitive' people. Then, for reasons not unlike those which drive environmentalists, the views of native people became symbolic ways of speaking about nature – symbolic because as everyone knew (everyone except the anthropologist's natives, that is) one cannot really have a relationship with a rock, animals and plants do not have intentions, the Earth is not inhabited by spirits nor does it have a spirit of its own. Could it be the case, then, that the language of the radical ecologies is also a symbolic, metaphorical way of speaking about nature? It, no doubt, could, but radical environmentalists themselves rule out such a possibility. They claim to 'know full well the meaning and use of metaphor'; and because they do know, they use metaphors 'only if they … are disclosing real dimensions of [other] creatures' subjectivity'. Take trees, for example. Does it make sense to say that they have hands? Apparently 'it does make sense to speak of hands metaphorically in relation to trees, since trees do, in fact, in their own fashion, respond to their Creator, both with deep groans of longing and pain and with songs of praise' (Walsh *et al.* 1996: 426–27). Since trees too respond to their Creator, it makes sense to speak of their branches as hands. The metaphor helps us to understand this 'real' aspect of the trees' 'subjectivity'. What environmentalists say, then, may not always be literal but nonetheless there is a fundamental difference between them and the anthropologist's natives. Environmentalists are well aware that they are using metaphors. They are also well aware that they are doing so to express a truth that is literal and true. The anthropologist's natives, by contrast, are aware of neither.

Anthropologists sympathetic to radical environmentalism are themselves beginning to question the wisdom of symbolic interpretation. They have now come to see that 'this kind of interpretation can contradict what people clearly know to be true' (Milton 2002: 46; see also Ingold 2000). Anthropologists, of course, have for the most part contradicted what natives 'clearly knew to be true', particularly about nature, without being excessively concerned. If symbolic interpretation can no longer be allowed to contradict the truth, this is only because anthropologists are now dealing with a different kind of people – their own. Take the truth about whales, for example. 'Anti-whaling campaigners do not construct whales as symbolic persons, they know that whales are persons because they perceive them as such. If they did not, they would not object so strongly to their being killed' (Milton 2002: 46). A tautological statement, one might say, and an epistemology of sorts. But perhaps the rules of logic and the standards of truth by which anthropologists routinely judge other people can be allowed to relax somewhat when they judge themselves. If 'people' perceive whales as persons, then that is exactly what whales are. If they think that trees sing and praise their creator, then that is exactly what trees do. Rocks *can* become conversational partners and friends and the Earth *has* a spirit to be invoked for enlightenment and guidance. All these things are real, not symbolic and metaphorical. They are real because this is how 'people' perceive them.

We have come a long way since the 1950s and what was then known with as much conviction and certainty to be true. A short time as historical change goes, but a radical change nonetheless, nothing short of a reversal in the meaning of the world. 'Man' the master of 'nature' has disappeared from the scene and if he still exists, he must be leading a subterranean life. The new being that has emerged, the human being, is by all accounts more cautious, more considerate and far less considerable – indeed, by some accounts, a mere drop in the ocean, a tiny, ephemeral and insignificant being. It would seem that there was little choice in the matter. 'Our only hope for ... renewal is our awakening to the realization that the Earth is primary and that humans are derivative' (Berry 1996: 411). Such is the new 'anthropology'. 'Nature' itself has been displaced by nature, the machine by the living organism, the object to be mastered by the subject to be protected, respected and cared for, the material and the mundane by the en-spirited and the sacred. Nature is sacred because 'people' perceive it in this way – because its 'value is not based on reason but is experienced directly, through the senses and, when necessary, asserted dogmatically' (Milton 1999: 440). Such is the new '(meta) physics'. We have come a long way from the 'anthropology' and the 'physics' of the modernist paradigm, which were also known to be true and were also 'asserted dogmatically when necessary' – as in the colonies, for instance. Yet this is not quite the end of the road. There is one last actor involved in this drama whose story must also be told if we are to achieve some sort of closure. The story of this actor runs parallel to the story of 'nature'-cum-nature and 'man'-cum-'human being' and is inextricably associated with it. It is the story of the anthropologist's 'savage'-cum-symbolic persona, of the United Nations expert's 'traditional and underdeveloped man'. It should come as no surprise that in the last act of this drama, this Other persona has been assigned a new role and displays a different identity.

'Our Debt to the Savage'

When the physical world becomes 'nature' – a disenchanted domain of utility and danger to be mastered and taken advantage of – 'man' has already appeared on the scene and has set the stage for what is to come. This is not just any man. It is 'man' the measure of all things, the centre of the physical world and its only relevant Subject. It is also 'man' the measure of all other men, the centre of the social world and *its* only relevant Subject – the 'man' who grounds everything, including, paradoxically and no doubt fatefully, himself. This, in short, is 'European man'. In the meantime, all those people in the rest of the world who did not think of themselves as 'man' and had therefore little need either for a 'physics' or an 'anthropology' – 'no nature, no culture', says Marilyn Strathern (1980) – had to be classified and dealt with. And they were. The death of 'nature' and its master and the birth of the human being and the environment generated the need for another reshuffle of the people with 'no nature and no culture', another round of classification, ordering, positioning. And once again, this is exactly what happened. It had to happen. Whether they know it or not, whether they like it or not, these people have a vital role to play in any change in the order of the world and the world order. They are key building blocks, which is to say, the blocks that bear most of the burden. Not that this onerous task has gone unrecognised or unappreciated. The architects of the world order have always expressed their gratitude – from the Victorian anthropologist to the contemporary environmentalist.

'Our gratitude is due to the nameless and forgotten toilers, whose patient thought and active exertions have largely made us what we are', wrote Sir James Frazer (1950 [1922]: 307) in the *Golden Bough* and devoted an entire chapter, 'Our Debt to the Savage', to explain how much we owe. Almost a century later, Klaus Töpfer, Executive Director of the United Nations Environment Programme, expressed similar sentiments:

> The very origins of environmental conservation lie buried in ancient cultures found throughout the world. Modern environmental movements express various ideologies of these *original* belief systems, yet do not always realize their debt to their forebears, nor towards those who *still embody* these ideals. Learning and respecting the ways of today's indigenous and traditional peoples, and integrating them into environmental and developmental considerations, will prove indispensable for the survival of diversity. (UN 1999: xi; my emphases)

Two expressions of gratitude, then, almost a century apart. Between them lie two radically different visions of the world and within these visions two very different personas – the 'savage' who did what he did because he was ignorant and innocent and did not know any better and the 'indigenous and traditional peoples' who have always known better than those who had underestimated them so much. The story of the 'savage' is well known to anthropologists but it is worth retelling, even if in broad outline, as a way of highlighting the contrast with the environmentalist image of native populations currently in fashion. Let us, then, return to Sir James Frazer to examine his reasons for being so grateful to 'the savage'.

'A savage' Frazer points out, 'hardly conceives the distinction commonly drawn by more advanced peoples between the natural and the supernatural'. The distinction, it should be noted, was definitively and irrevocably drawn by the 'more advanced peoples' in the eighteenth century – by David Hume (1977 [1748]: 76), for example, who, as we have seen, argued passionately against the possibility of miracles because they constituted a violation of the laws of nature:

'As a firm and unalterable experience has established these laws, the proof against a miracle ... is as entire as any argument from experience can possibly be imagined' – such was Hume's certainty. To the 'savage', Frazer (1950 [1922]: 11) goes on to say, 'the world is to a great extent worked by supernatural agents, that is, by personal beings acting on impulses and motives like his own, liable like him to be moved by appeals to their pity, their hopes, and their fears'. Why 'the savage' failed to make the crucial distinction between the natural and the supernatural had been explained in the new science of anthropology by E.B. Tylor (1874: 428) a few decades earlier. 'It seems as though thinking men, as yet at a low level of culture, were deeply impressed by two groups of biological problems'. The first problem was 'the difference between a living body and a dead one', the second, 'the human shapes which appear in dreams and visions'. To explain these problems, the early 'thinking men' developed the idea of the soul or spirit and, being 'at a low level of culture' and not knowing any better, went on to attribute the same substance to other biological beings.

> The sense of absolute psychical distinction between man and beast, so prevalent in the civilized world, is hardly to be found among the lower races. Men to whom the cries of beasts and birds seem like human language, and their actions guided as it were by human thought, logically enough allow the existence of souls to beasts, birds, and reptiles, as to men. (1874: 469)

A 'logical enough' inference perhaps, but the wrong one nonetheless – wrong, that is, until the rise of environmentalism for which the 'absolute' psychical distinction between 'man' and 'beast' is no longer so absolute. Here is another error committed by the 'lower races'. 'Plants, partaking with animals the phenomena of life and death and sickness, not unnaturally have some kind of soul ascribed to them' (Tylor 1874: 474). Not unnaturally but not correctly either. Yet, Tylor goes on to say, this was not all. '[At] the primitive stage of thought ... personality and life are ascribed not to men and beats only, but [also] to things. ... What we call inanimate objects – rivers, stones, trees, weapons, and so forth – are treated as living intelligent beings' (1874: 477). Everything, then, was a person, had a soul and was treated accordingly. Take trees, for instance – since, as we have seen, for many environmentalists trees 'respond to their Creator, both with deep groans of longing and pain and with songs of praise'. As Frazer (1950 [1922]: 129–30) points out, 'formerly the Indians [of the upper Missouri] considered it wrong to fell [cottonwood]'. Indeed, 'till lately some of the more credulous old men declared that many of the misfortunes of their people were caused by this modern disregard for the rights of the living cottonwood'; the Wanika of eastern Africa 'fancy' that all trees have a spirit and regard '"the destruction of a cocoa-nut tree [in particular] ... as equivalent to matricide"', while the Ojebways '"very seldom cut down green or living trees, from the idea that it puts them to pain"'. Such was the credulity and innocence of these people. Or take the way they treated animals. 'On the principles of his rude philosophy the primitive hunter who slays an animal believes himself exposed to the vengeance … of its disembodied spirit'. He therefore 'makes it a rule to spare the life of those animals which he has no pressing motive for killing'. Yet often animals must be killed for food, in which case the primitive hunter 'is forced to overcome his superstitious scruples and take the life of the beast'. But then he does everything possible to appease his victims. 'Even in the act of killing them he testifies his respect for them, endeavours to excuse or even conceal his share in procuring their death' (Frazer 1950 [1922]: 601, 603). Not a set of customs to be criticised as such perhaps, but nonetheless indicative of the 'rude philosophy' responsible for the 'primitive' condition of such people.

And yet, Frazer (1950 [1922]: 306–307) points out, it would be a mistake to dismiss the 'savage' completely. And it would be an even greater mistake to assume that his 'errors were … wilful extravagances or the ravings of insanity'. As Tylor himself argued in his own analysis of savage folly, the 'savage's' reasoning was logical enough and his inferences not 'unnatural'. Frazer concurs. Savage philosophy may be 'crude and false' but at least it has the merit of 'logical consistency'. Rather, the flaw of the system, though a 'fatal one … lies not in its reasoning, but in its premises'. It had to lie in the premises and not in the reasoning. If 'savages' were, indeed, 'insane', how were Tylor and Frazer to explain the emergence of sanity and reason in the 'civilised world'?[7] 'Contempt and ridicule or abhorrence and denunciation are too often the only recognition vouchsafed to the savage and his ways', Frazer complains. Yet to 'stigmatise' the savage in this way 'would be ungrateful as well as unphilosophical'. It would be ungrateful because 'we stand upon the foundations reared by the generations that have gone before'. It would be 'unphilosophical' because it would undermine the 'philosophy' to which Frazer, Tylor and other luminaries before and after ascribed. It would spoil the grand vision that they carefully put together, which at once explained and legitimised the order of the world and the world order: 'man' struggling with 'nature' over thousands of years, to achieve definitive victory only during the last few centuries in the guise of 'European man'. Frazer, Tylor and their contemporaries had every reason to be grateful to 'the savage'. By being 'ignorant', and in his ignorance, he made it possible for Europeans to imagine themselves as the 'more advanced people' and to treat the rest of the world accordingly.

Many things have changed in anthropology since the time of Tylor and Frazer, including the discipline's central problematic. Beginning with Malinowski in England and Boas in the United States, the main issue was no longer defence of the 'psychic unity of mankind'. Not that racism disappeared. Rather, the view of native societies as culturally inferior, which Tylor and Frazer took for granted, became less palatable and eventually unacceptable. It became 'ethnocentric'. One thing that did not change, however, was the view of the physical world as a disenchanted domain, the belief, that is, in the categorical distinction between the natural and the supernatural. Hence, the new ethnological problem: how to explain the native population's apparent disregard of this distinction without resorting to ethnocentrism. As I argue elsewhere in greater detail (Argyrou 2002), the anthropologist's efforts to deal with this problem produced a different kind of native – not ignorant, superstitious, or innocent but 'symbolic'. According to this view, native disregard of the distinction between the natural and the supernatural is not literal but metaphorical. Natives are well aware that the two realms are separate; they demonstrate this awareness in many of their practical pursuits. The apparent conflation of the two realms is therefore to be read as a symbolic statement. Its primary aim is not to manipulate the natural through the supernatural but to make a statement about the social and the cultural: to explain misfortune (Evans-Pritchard), for instance, to classify the sensible world and resolve its contradictions (Lévi-Strauss), to come to terms with conflicts in the conceptual and social order (Douglas), or to maintain a meaningful cosmos (Geertz). This is not to say that native populations recognised themselves in the symbolic persona that the anthropologist constructed for them. Witchcraft, for instance, which for Evans-Pritchard was the means with which the Azande explained misfortune, was for the Azande themselves what it had always been – a force located in the body of the witch that could fly through the air to attack the witch's enemies and could be clearly seen, especially at night, as a flash of light. This is what the Azande 'clearly knew to be true'. That symbolic interpretation contradicted this truth was neither here nor there. The important thing was what the anthropologist knew to be true. As Geertz (1993: 58) admirably put the matter, 'the trick is to figure out what the devil they [natives] think they are up to'. For what they think they are up to, and what they 'really' are up to are two different things – and the latter is known only by the anthropologist.

Perhaps this is still the 'trick'. Indeed, to a large extent, it has to be. If what the anthropologist's natives thought of as the truth were accepted as such, anthropologists would be reduced to mere translators of exotic languages. The anthropological 'trick' is to posit an exotic culture full of symbolism that can be 'thickly' described. Yet as we have seen, things are changing in a direction that does not favour the anthropological practice of 'thick' description. Environmentalists, including many anthropologists sympathetic to the environmentalist cause, now find much in 'exotic' societies that requires no cultural translation. What native populations know can now be understood directly and, as the claim has it, in the same way in which they themselves mean and understand it. It goes without saying that for environmentalists and anthropologists of the same persuasion, native populations are no longer 'symbolic' personas. On the contrary, they mean what they say and much of what they say is true. Hence, the 'trick' now is to 'figure out' how to use what they say to undo the damage caused during the long centuries that 'we' thought otherwise about what they said and what they meant. How this reversal in the perception of native populations has come about is difficult to say. If the following example is anything to go by however, it seems that it all began without prior thought or design – because, for instance, a 'vague affinity' that was, somehow, already there was strengthened by a chance visit.

In 1971, a group of American and Canadian counter-culture activists living in Vancouver hired an old boat and set sail for Amchitka, one of the Aleutian Islands off the coast of Alaska, to protest against American nuclear testing there. The boat, *Phyllis Cormack*, was for this particular journey re-christened 'Greenpeace' after the name by which the group of activists had recently decided to call themselves.[8] As the chroniclers of the journey point out, on the way to Amchitka, the crew made a stop at an Indian village on Cormorant Island and received gifts and blessings from the local population. It seemed, recalled Robert Hunter, one of the activists, 'to strengthen the "vague affinity" most of the crew felt with the Indians' (Brown and May 1989: 12–13). This 'vague affinity' was to become clearer and develop further when the activists resumed their journey. 'As the *Cormack* pulled away from Cormorant Island, Hunter passed around' his copy of a small book of Indian myths and legends. It was entitled 'Warriors of the Rainbow'.

> The book contained a 200-year-old prophecy that seemed particularly relevant to the men on board the *Cormack*. There would come a time, predicted an old Cree woman named Eyes of Fire, when the earth would be ravaged of its resources, the sea blackened, the streams poisoned, the deer dropping dead in their tracks. Just before it was too late, the Indian would regain his spirit and teach the white man reverence for the earth, banding together with him to become Warriors of the Rainbow.

Not everyone on board found the book and its message particularly inspiring or relevant. '"Predictably", [Hunter] wrote later, "the older men were less impressed than the youngsters. But rainbows *did* appear several times the following day and it all *did* seem somehow magical"'. 'Predictably', one presumes, because the older men on board were far too entrenched in the empiricism of modernist paradigm to take Indian myths and prophecies seriously – counter-culturalists and environmentalists or not, and even if everything seemed 'somehow magical' during the journey.

Such, then, may have been the uncertain beginnings of the transformation of anthropology's native – the-innocent-savage-cum-symbolic-persona – one of the many instances of how environmentalists developed an 'affinity' with native populations no doubt, and not necessarily the first or the most important one. The point in this story, in any case, is more general. It is about how the stories of native populations, such as the Cree woman's prophecy, began to appear 'particularly relevant', how, that is, a set of cultural assumptions – in this case the assumptions of the counter-culture of the 1960s and 1970s – rendered certain things visible for the first time or, at any rate, visible in a new 'light', how they generated a domain of meaning and significance that previously did not exist or was only 'vaguely' understood. This is to say also that the transformation of native populations was not an act of will, intention or planning. It was rather, as much as the earlier constructs by which native populations have come to be known at different points in time – indeed, as any construct – a discovery. The cultural context that was beginning to emerge in the early 1970s rendered the current image of native populations immanent, a virtual reality waiting to be discovered, articulated, given definite shape and form. Because this image was not an act of will or intention, it cannot be a cultural construct for environmentalists. It appears to them as an objective reality – a 'fact'.

As the assumptions about the nature of nature and the nature of humanity became more widespread, as they became more relevant and began to consolidate themselves into a clearly articulated 'physics' and 'anthropology', so did the new image of native populations in the environmentalist imagination. The uncertain beginnings and vague sentiments were replaced by an unprecedented recognition of native populations that surpasses by far the recognition accorded to the 'noble savage' by the European Romantics of the eighteenth century. Not only was the 'ecological' way of life and 'environmental ethic' of native populations acknowledged, but also their rights, territories and newly acquired status were defended passionately both in national and international contexts – all of which eventually became relevant and meaningful to, and received the stamp of approval of, the most important of international bureaucracies: the United Nations.

There is no mention of 'indigenous and traditional peoples' in the reports of the early United Nations conferences on the environment, let alone any suggestion that they may be in possession of ecological knowledge and experience that could be used by the rest of the world to deal with its environmental problems. In the early 1970s, the experts involved in environmental issues were not likely to be impressed by such a proposition – less likely, one suspects, than the older men on board the '*Phyllis Cormack-cum-Greenpeace*'. In one such early report we read: 'We are not sleepwalkers or sheep. If men had not hitherto realized the extent of their planetary interdependence, it was in part at least because, in clear, precise, and scientific fact, it did not exist'. Had it existed, 'men' would have known about it. Now that science has formulated the issue in 'clear' and 'precise' terms, they have become fully aware of it. Not only do they now know, but they are also already thinking ahead about possible solutions. 'The new insights of our fundamental condition can also become the insights of our survival'. Indeed, the happy coincidence is that 'we may be learning just in time' (Ward and Dubos 1972: 290). Such was, and no doubt still is, the uncompromising spirit of humanism. If a problem comes to their attention, 'men' deal with it quickly and effectively. They have the will, the knowledge and the power to do so. They are neither 'sleepwalkers' and daydreamers nor passive and docile like 'sheep'.

By the mid 1980s the proposition that native populations may have retained a certain 'ecological wisdom' that, to its detriment, the rest of the world had long forgotten had gained enough ground to circulate widely as a serious and legitimate statement reproducible even by the experts commissioned by the United Nations.

> The isolation of many [native populations] has meant the preservation of a traditional way of life in close harmony with nature. Their very survival has depended on their ecological awareness and adaptation. ... These communities are the repositories of vast accumulations of traditional knowledge and experience that links humanity with its ancient origins. Their disappearance is a loss for the larger society, which could learn a great deal from their traditional skills in sustainably managing very complex ecological systems. (WCED 1987: 114–15)

A few years later a joint publication of the World Conservation Union (IUCN), the United Nations Environment Programme and the World Wide Fund for Nature returns to the people who 'link humanity with its ancient origins' to reiterate the very same message for the benefit of those who had long severed the link.

> Respect for other forms of life is easiest in those cultures and societies that emphasize that humanity is both apart from and a part of nature. It is most evident in those communities whose lives are lived in close contact with nature, and whose traditions of care for it endures. This is the basis of the special contribution that indigenous peoples can make to the rediscovery of sustainable living by the world community. (IUCN *et al.* 1991: 14)

The following year the 'world community' gathered in Rio with precisely such an aim in mind – to 'rediscover sustainable living'. And it declared its full support to the proposition that 'indigenous and traditional peoples' were the very people equipped with the ethos, knowledge, experience and skill to assist humanity in this journey of rediscovery' – which, no doubt, is also a journey of self-discovery. Principle 22 of the 'Earth Summit' states:

> Indigenous people and their communities and other local communities have a vital role to play in environmental management and development because of their knowledge and traditional practices. States should recognize and duly support their identity, culture and interests and enable their effective participation in the achievement of sustainable development. (UN 1992: 11)

We have come a long way both from the anthropologist's savage-cum-symbolic-persona and the expert's 'underdeveloped man'. Yet this is still not the last word about 'indigenous and traditional peoples'. There is more – more polemical, more militant, more subversive of modernist assumptions, a discourse that attempts a sort of 'status reversal' between the West and its Others, which, as we shall see in the next chapter, is also a source of anxiety to the apologists of the modernist paradigm and the focus of their critique of environmentalism. The last and more subversive word is, of course, the discourse of radical environmentalism. Let us, then, turn briefly to a paradigmatic example, a comparison between American Indians and European colonisers. The former, according to the argument, lived on the continent for thousands of years,

> without destroying, without polluting, without using up the living resources of the natural world. Somehow they had learned a secret that Europe had already lost … the secret of how to live in harmony with Mother Earth, to use what she offers without hurting her; the secret of receiving gratefully the gifts of the Great Spirit. (Hughes 1996: 131)

When the first Europeans arrived in the New World, Hughes goes on to say, 'the continent was so unspoiled that they mistook it for "wilderness" and regarded the Indians as "savages", wild denizens of a wild land. … But they were wrong. They could not have been more wrong'. They were wrong because:

> the condition of the New World as it met 'the eyes of discovery' was a testimonial to the ecological wisdom of the Indians. … It was the artefact of a civilization whose relationship to the living world was perceived by the Indians in terms that Europeans would not grasp at all. … If all the resources of modern anthropology, psychology, and linguistics are only now piecing together the picture of the ecology and culture of the Indians before Columbus, it is not surprising that the Europeans who first arrived did not understand what they were seeing. (1996: 132–33)

It is difficult to see the kind of 'status reversal' attempted in this extract without returning once again to Tylor and Frazer. Both the primary themes of radical environmentalist discourse – the nature of nature and the nature of civilisation – and the type of rhetoric employed are such that their full meaning and wider implications can become apparent only by contrasting them with a discourse preoccupied with similar themes and motivated by equally impassioned certainties. If we return to Tylor and Frazer, then, we see that what is attempted in this extract is more than a simple 'status reversal'. For Tylor and Frazer, the Europeans of their time were without doubt the most advanced people in the world. Yet they were not so advanced as to be unable to recognise and understand 'savage customs' – a claim implicit in Lévy-Bruhl (1926), for instance, and the argument that natives were not so much illogical as pre-logical, at a stage where logic had not been discovered yet (cf. Evans-Pritchard 1965). Both Tylor and Frazer pointed to examples of 'survivals' of such customs in European societies, particularly among the 'lower' classes and rural populations. By contrast, the claim in this extract is that American Indians were so far ahead of the Europeans who colonised the New World that the latter could not 'grasp at all' what they were witnessing. Being ignorant of the true nature of civilisation, they mistook it for savagery. It is only recently, according to Hughes, that Westerners are beginning to understand American Indian civilisation, and this after they have invested considerable intellectual resources and effort. And it is only recently that they are beginning to discover truths about nature that Native Americans have known for thousands of years.

> Ecologists in recent years have been trying to get people to think of the world in ecological terms: to see everything as connected to everything else, to see ourselves not as the rulers of the earth but as fellow citizens with all other forms of life.... American Indians would have recognized these ideas as soon as they were explained to them. Their philosophy was already ecological. (Hughes 1996: 144)

It is ironic that Westerners have come to understand Native American civilisation, even if only a little and belatedly, with the use of science – anthropology, psychology, linguistics and ecology. For was it not science, after all, or at any rate, the 'scientific spirit' that, as environmentalists themselves argue, blinded Europeans to the non-scientific but far superior American Indian civilisation? The irony highlights the paradox we have already encountered and which Harriet-Jones captures quite graphically in the aphorism that radical environmentalists operate in a space 'between science and shamanism'. It also raises important questions about the precise nature of the current valorisation of native populations that need to be addressed.

Although examples of both United Nations and radical environmentalist discourses about native populations abound, enough has been said to highlight the current image and to allow a contrast with the positivist image that was dominant well into the second half of the twentieth century. What for the Victorian anthropologist was so obviously a 'crude' and 'false' natural 'philosophy' turns out to be 'ecological awareness' and an ethical system of 'respect' and 'care' for other forms of life. What until recently was little more than ignorance and superstition now emerges as nothing less than sound empirical 'knowledge', 'skill' and 'experience' – so sound and reliable, in fact, as to have a 'vital role' to play in the 'environmental management' of the world. What for the experts reporting to the United Nations in the 1950s and 1960s were 'ancient philosophies [that] have to be scrapped [and] old social institutions [that] have to disintegrate' is for the experts reporting in the 1980s and 1990s 'ways of life' in their own right, 'identities' and 'cultures' that States should 'recognize and duly support' – indeed, by some accounts, a civilisation far superior to anything Westerners have ever managed to produce. What is one to make of this remarkable change of heart?

There is something peculiar about these discourses. Although they emphasise the knowledge, experience and skills of native populations and insist that they are critical tools in any effort to deal with the 'ecological crisis', there is hardly any mention of what these tools might be. Nor is there an explanation as to how exactly they could be used. It could be the case, of course, that these are programmatic statements and that the actual mechanics of transferring native expertise and applying it to the environmental problems at hand are yet to be decided. But it could also be the case that this silence has something to do with the incompatibility of 'science and shamanism'. Anthropologists concerned with native populations insist that the latter treat their physical surroundings as an extension of society, which is to say, animals, plants and things as persons. This is as true of nineteenth-century anthropologists who, as we have seen, explained the practice as a manifestation of ignorance and innocence, as of many contemporary anthropologists who have abandoned symbolic interpretation in defence of nature and the native populations' treatment of it (e.g. Descola 1996 and Ingold 2000). The current 'physics', by contrast, as is reflected in the science of ecology, presents nature as a fragile domain of 'very complex ecological systems', which needs to be 'managed' in a 'sustainable' fashion. How, then, can the knowledge, experience and skills of native populations, which have been developed to allow them to function 'in the society of nature' (Descola 1996) be used to manage 'ecosystems'? Questions of this sort have been raised by several anthropologists who are sympathetic to the environmentalist cause but find unconvincing the idea that native populations possess the kind of ecological knowledge attributed to them. Milton, for instance, refers to this idea as 'the myth of ecological wisdom' (see also Ellen 1986). As she points out, 'non-industrial peoples do not think like environmentalists. Some of them may live their lives in ways that are environmentally sound, but ecological balance, where it exists, is an incidental consequence of human activities and other factors, rather than being an ideal or a goal that is actively pursued' (Milton 1996: 113).

But perhaps the environmentalist statements about the knowledge, experience and skills of native populations are not meant to be read so literally. Perhaps the role that they are expected to play in environmental management is more symbolic than practical. It could be the case, for instance, that what is 'vital' is not so much their practical knowledge and skills as what they are meant to represent, namely, the ideal of living in 'harmony with nature'. It could be the case that the 'special contribution' native populations are expected to make is to act as an inspirational symbol of one sort or another. If so, this is to say also that, to use a Lévi-Straussian phrase, native populations are 'good to think with'. At the most immediate level, they are good to think with about questions of individual and collective identity as well as the current distribution of power: what caused the 'environmental crisis' (an alienated, misinformed, spiritually impoverished culture), who is to be criticised and held accountable for it (the improvident steward, 'man' in his arrogance, patriarchy, capitalism), what is to be done about it (change the patterns of thought and structures of action). At another level, they are good to think with about broader ontological issues: the 'mysteriousness and open-endedness of existence', the 'radically unknown character of the future', 'humankind's place in creation' – concerns that seem once again to be in 'the focus of European thought' and the subject of 'protracted reflection'. Native populations, in short, may be good to think with and speak about 'us' rather than about 'them'. If that is the case, the image of native populations as repositories of ecological wisdom may well be the latest 'ethnographic allegory' (Clifford 1986) to have emerge in the West – in form the reverse of the ethnographic allegory produced by Victorian anthropologists but in structure exactly the same. This is to say also that environmentalists have as much of a reason to be grateful to the 'savage' as Tylor and Frazer. Native populations are once again used as key building blocks in the latest Western construct – the environmentalist vision of the world. What is more, they themselves emerge as the kind of willing 'toilers' that Frazer had in mind: 'We are trying to save the knowledge that the forests and this planet are alive, to give it back to you who have lost the understanding', says Bepkororoti Paiakan, a chief of the Kayapó people of Brazil. 'We must uphold the basic rights of indigenous and traditional peoples to land, territory, knowledge and traditional resources', says the environmentalist who quotes him. 'And we must discover', he goes on to add, 'how the balance sheet of economic and utilitarian policies can be countered by the "sacred balance" expressed by such people' (Posey 1998: 103). Upholding the basic rights of 'indigenous and traditional peoples' is the least environmentalists can do. Their vision of the world depends on 'such people' being 'indigenous' and 'traditional'.

Notes

1. The 'unofficial report' of the Conference was published as *Only One Earth: The Care and Maintenance of a Small Planet* (Ward and Dubos 1972).
2. This is the title of a subsequent anthropological paper on the environment not related to the United Nations Conference (Bird-David 1990).
3. For statements by religious leaders, including the Pope, supportive of the environmentalist cause, see Gottlieb (1996).
4. This is the title of a radical environmental 'manual' that consists of essays, 'invocations' and prayers for the healing of 'Mother Earth' (Gottlieb 1996).
5. The main radical ecologies are 'deep ecology', a term coined by the Norwegian philosopher and radical environmentalist Arne Naess, 'social ecology' and 'ecological feminism' or 'ecofeminism'. For an overview, see Zimmerman (1994) and Zimmerman *et al.* (1998).
6. As one commentator put it, radical environmentalists operate in a space 'between science and shamanism' (Harries-Jones 1993).
7. On this issue, see Stocking's (1987) excellent discussion.
8. The group initially called itself 'The Don't Make A Wave Committee'. Yet, as one of the founders put it, 'it was a lot of words and didn't mean much' (Brown and May 1989: 9).

3 The Logic of the Same

The Phenomenology of Change

In dismissing the modernist 'physics' and 'anthropology' as a grand historical mistake that we cannot afford to repeat, in rejecting both 'man' and 'nature' in favour of the 'human being' and the 'fragile planet' or more controversially 'sacred earth', environmentalism constructs itself as a radical rupture with modernity. As will become apparent below, critics of environmentalism depict it in a similar manner, even if for quite different reasons. Common sense too suggests that there is a large gap between modernity's philosophy of nature and culture and the environmentalist perception of the world; and that however rhetorical and exaggerated the claims of environmentalists and their critics might be, there can be no doubt that the world is undergoing significant change. Everything, then, seems to point in the same direction and at the same thing – a radical break with long-held assumptions and expectations, a fundamental transformation of long-established attitudes and practices, indeed, a reversal of our view of the world and our place in it. Such is the phenomenology of the change that is currently upon us.

My aim in this chapter is to raise the sort of question on which all social analysis depends, namely, the question that concerns itself with what lies behind appearances. To raise this question is not to doubt the obvious – the phenomenon of change as a phenomenon. There is no doubt that 'nature' has been transformed into nature and 'man' reduced to the size of the 'human being'. It is, rather, to examine the conditions of possibility of the new 'physics' and 'anthropology', those structures – inextricably logical, epistemological and ontological – that may account for the emergence of the environmentalist vision of the world. My aim, in short, is to search for the cultural logic of environmentalism: 'cultural' because it defines a particular historical period in a particular part of the world – the West; and a 'logic' because although it may not necessarily be understood as a system, it does produce systematic and predictable results, which is not to say inevitable. The general argument, as it will develop in the present chapter and the next one, is that at a deeper and more profound level than the phenomenological, environmentalism reproduces the Same of the modernist paradigm. It operates on the basis of the modernist logic of the Same, the logic that strives to efface the divisions of the Whole and maintain its imaginary unity. Environmentalism does not simply reproduce this logic. It takes it to its logical and ontological extreme. It strives to efface, which is to say prove groundless, the last and grandest of all modernist divides of the Whole – the division between humanity and nature.

Before the question as to what may lie behind appearances can be raised, however, it is necessary to complete the task of determining the extent of the phenomenological rupture by contrasting the environmentalist position with that of its critics. I shall begin by restating briefly the environmentalist case in its least controversial and more widely accepted version – the version, that is, which, as we have seen, radical environmentalists dismissively call 'mainstream' environmentalism or 'conservationism'. Let us, then, turn briefly to an individual who is authorised to speak on behalf of the entire world or, at any rate, on behalf of the representatives of the entire world – and hence an individual also who must speak if not dispassionately, certainly well within the bounds of the moderate and the sensible. Let us turn to Maurice F. Strong, Secretary-General of the United Nations Conference on Environment and Development, and his Foreword to 'Agenda 21: Programme of Action for Sustainable Development'.

> Humanity today is in the midst of a profound *civilizational change.* There are signs of it everywhere ... *exhilarating,* uplifting signs. ... Industrialists, economists, financiers, engineers, scientists – those who, in truth, hold the levers of economic power and change – have joined the constituency of earnest environmentalists in a commitment to the fulfilment of the hopes and aspirations engendered by Rio. In short, the movement to turn the world from its *self-consumptive* course to one of *renewal* and sustenance has unmistakably spread from the grass roots to the brass roots. (UN 1992: 1; my emphases)

Anthropologists would readily recognise several ritualistic and mythical themes in this passage – moral community, communitas, regeneration and rebirth. Let us note also the connotations of disease in the term 'self-consumptive course' and the way in which the change that is currently upon is depicted – nothing less than a change in 'civilisation'. Rhetorical claims, one might say, demanded by the circumstances. No doubt, but this is hardly a reason to take them lightly. Similar themes emerge in statements made by politicians both at the Rio Conference and on other occasions.

'Our generation has seen our planet from space', said the Prime Minister of Canada in his speech at the Earth Summit in Rio. Have we learnt anything by taking such a long view of the world? 'We know its beauty and we understand our fragility. We know that nature is part of us as we are part of nature' (UN 1993: 72). What we have learnt, then, is the truth about the nature of nature and the extent of our misconceptions. Nature is beautiful – not a pool of raw materials and resources; and it is fragile – not the refractory domain that we have imagined it to be. Nature is both beautiful and fragile, and we are very much part of it, not its masters. 'We are in fact', the French President asserted at the same summit, 'the first generations, some 3 million years after our distant ancestors emerged, to acquire awareness of the physical laws which govern us'. What are these laws that we have taken so long to discover? 'The first', said François Mitterrand, 'is that the Earth is a *living* system whose parts are interdependent, and therefore that the destinies of all species – human, animal, plant – are connected'. The second law 'tells us that the resources of the Earth are limited' and the third 'that humankind cannot be separated from nature, for it is part of nature itself *just like* water, trees, the wind or the depths of the seas' (1993: 193; my emphases). We are the first to know and the first to carry the burden of this knowledge. 'One day they will say to us: You knew all that and what did you do? That is the true topic of our Conference'. What is to be done, then, now that we know 'all that'? 'Our duty', said the French President in the manner of a humble peasant who works the soil and gives thanks to the earth, 'is to see to it that the Earth which sustains us is both our dwelling and our garden. Our shelter and our food' (1993: 193). Such is our duty and we must prepare ourselves for it. For 'we are ... going to experience a change of the magnitude of those of the neolithic age and the beginning of the industrial age'. What we will experience will be nothing less than the third most important revolution in the history of humanity.

In a different forum but in a similar vein, even if not with the use of such grandiose analogies, Al Gore, United States senator and former Vice-President, depicts the struggle to save the environment as a continuation of the struggles against two other modernist afflictions, Nazism and Communism. 'It is not merely in the service of analogy that I have referred so often to the struggles against Nazi and communist totalitarianism, because [sic] I believe that the emerging effort to save the environment is a continuation of these struggles' (Gore 1992: 275). Modernity, according to Gore, is a 'dysfunctional civilization'. Beyond its totalitarian tendencies, it suffers from an inauthentic culture. We live in 'a false world of plastic flowers and AstroTurf, air-conditioning and fluorescent lights, windows that don't open ... days when we don't know whether it has rained or not ... sleepy hearts jump-started with caffeine, alcohol, drugs and illusions', a culture, in short, which is a poor substitute for 'direct experience with real life' (1992: 232). We have been alienated from nature – this is the primary source of our current problems. Instead of confronting the pain that alienation causes, we suppress it and 'search insatiably for artificial substitutes to replace the experience of communion with [nature]' (1992: 231). The more substitutes we create, the more we destroy the earth. 'Our civilization is, in effect, addicted to the consumption of the earth'. The addiction 'distracts us from the pain of what we have lost: a direct experience of ... the vividness, vibrancy, and aliveness of the rest of the natural world' (1992: 220). Such is the predicament that we have brought upon ourselves. But perhaps not everything is doom and gloom, perhaps there is still hope. 'Ladies and Gentlemen', said François Mitterrand closing his speech at the Earth Summit, 'I urge you to make the name of Rio 1992 stand for hope' (UN 1993: 196).

Yet not everyone is applauding. To begin with, there are those, among them the odd 'sceptical environmentalist' (Lomborg 2001), who are not convinced that the world is facing ecological problems of the magnitude claimed by environmentalists. There are also those apologists of the modernist paradigm who are seriously concerned about what 'turning the world round' means, what it would involve and where such a reversal would eventually take us.

Let us first examine briefly the objections about the extent of the 'environmental crisis'. 'It is now generally accepted', writes Lovelock in *Gaia: A New Look at Life on Earth*, 'that man's industrial activities are fouling the nest and pose a threat to the total life of the planet which grows more ominous every year'. Is this really the case? Lovelock does not think so. 'It may be that ... our technology will in the end prove destructive and painful for our own species, but the evidence for accepting that industrial activity either at the present level or in the immediate future may endanger the life of Gaia as a whole, is very weak indeed' (2000: 1001–101). Lovelock's 'Gaia hypothesis' is a mixed blessing for environmentalists. On the one hand, it depicts the earth as a living organism, a view that environmentalists themselves share. On the other, it argues that the earth can take care of itself, which makes environmentalism more or less redundant or, at any rate, undermines the claim that what is at stake in environmental degradation is life on earth writ large. As we have seen, for Lovelock 'Gaia' is a self-regulating system that ever since life emerged on the planet maintains relatively constant conditions under which life can flourish. 'Whenever natural disasters occur ... there is turmoil among species'. Over time, however, 'a new ecosystem comfortable with the new environment emerges and is populated by new species of organisms' (2000: 102). Species can and do get extinct but the self-regulating mechanism of 'Gaia' ensures that life itself is not affected. For Lovelock (2000: 103), 'the very concept of pollution is anthropocentric and it may be irrelevant in the Gaian context' – irrelevant, that is, for life writ large. As for the environmental hazards that we and other species may be facing *as species*, these can be dealt with 'by retaining but modifying technology than by a reactionary "back to nature" campaign' (2000: 110).

The extent of the 'environmental crisis' and the imputed dangers posed by modernity, both literal and symbolic, have recently been brought under rigorous questioning by a 'reformed' Danish environmentalist, Bjørn Lomborg. Lomborg summarises his argument against what he calls the 'litany' of environmental disasters in the following way:

> We are not running out of energy or natural resources. There will be more and more food per head of the world's population ... Global warming, though its size and

future projections are rather unrealistically pessimistic, is almost certainly taking place, but the typical cure of early and radical fossil fuel cutbacks is way worse than the original affliction ... Nor will we lose 25–50 percent of all species in our lifetime – in fact we are losing probably 0.7 percent. Acid rain does not kill forests, and the air and water around us are becoming less polluted. (2001: 4)

This summary dismissal of environmentalist claims is supported by statistical information spread throughout Lomborg's massive book in 173 figures and 9 tables. But Lomborg is equally dismissive of the less quantifiable, more ideological environmentalist criticisms of the modernist paradigm. Take Gore's argument, for instance, namely, that modernity is a 'dysfunctional civilization'. The claim, Lomborg argues, 'reveals both a *scary idealisation of our past* and an abysmal arrogance towards the developing countries of the world' (2001: 328; my emphases). For Lomborg the story of modernity is a story of success. '*We have more leisure time, greater security and fewer accidents, more education, more amenities, higher incomes, fewer starving, more food and a healthier and longer life.* This is the fantastic story of mankind, and to call such a civilization "dysfunctional" is quite simply immoral' (2001: 329; original emphases).

Environmentalists have not received Lomborg's views kindly. On his arrival in England to promote his book, *The Sunday Times* published an article under the title, 'Eco-heretic beset by hate campaign'. Lomborg, the paper says, 'dared to challenge the establishment view on climatic change' and is now paying a heavy price for it. '[He] has been subjected to a campaign of personal abuse, professional vilification and threats to his safety' (Leake 2002: 4).

In another critique, Richard North depicts environmentalism as an indulgence of 'the rich world ... in a dangerous idealism about the relations between human beings and nature', an idealism which is explicitly linked to ideas of secular salvation in a this-worldly Garden of Eden. Accordingly, he advises environmentalists to 'attempt a sounder reconciliation between [their] dreams and the realities of life' (North 1995: 5). The ecological disasters that environmentalists forecast may be a story that 'makes exciting broadcasting, sells newspapers, and attracts membership to campaign groups' but it is a myth, 'a largely invented idea'. And so is, according to North, another popular environmentalist idea: that indigenous populations live happily in the state of nature and have things to teach the rest of the world. 'In the neo-religion of the Greens, the tribal man and woman are Adam and Eve [and] their home is the garden of Eden ... before the serpent of greed makes people eat the apple of industrial development' (North 1995: 198). Here too reality is quite different. To make his point and to show what reality is really like, North asks those people who think 'it is wrong to force primitive people to accept' modernist culture to consider seriously 'what it would be like to live alongside a primitive culture which systematically denied its people ... elementary human freedoms' (1995: 241–42). North has already prepared his readers for this sort of reflection. Earlier in the book, he quotes from a study of 'the Stone-Age cultures' of Brazil, which claims to know what it is like to live alongside 'primitive' people. 'It was the continued threat of *slave-raiding, head-taking*, and in some cases *cannibalism*', says the study quoted by North, 'which held the non-riverine groups in their less favourable environment'.[1]

Idealistic or not, both the 'myth' of impending ecological disaster and the 'myth' of the 'noble savage' pose a serious threat to the 'rich world'. As it should already be apparent, what is at stake for North is far more than mere prosperity.

> Rationality, participatory democracy, freedom of worship, respect for the individual, and an interest in the natural world are the fundamental values of our society. We have struggled for centuries to find a system which allows individuals full expression, while binding them in a net of obligations which they do not find crippling. Our science and ambitions involve us in a continuous search for cheaper, more ecologically sound, better ways of making things, delivering services and treating people. These are encouraged in the real world of industrial society and constitute progress. The Greens have a modest role in helping that progress define itself, but were immodest when they thought that they could somehow transform our thinking. (1995: 251–52)

What is at stake, then, is the modernist culture itself: rationality, humanism, the individual and our fundamental freedoms. The Greens are 'immodest' in thinking that they can persuade us to change our way of life. They are 'immodest' in thinking that they can persuade us to give up what has been won after centuries of struggle. They may be 'immodest' but they are also dangerous. If they cannot persuade us through rational debate, they may attempt to force us to conform. Such fears are widespread among critics of environmentalism and give rise to charges of misanthropy, fundamentalism and 'ecofascism'.[2] 'The humanist era is being brought to a close', warns another apologist of the modernist paradigm in a trenchant critique of the radical ecologies (Ferry 1995: xix). This is 'the main objective for these new zealots of nature'. New zealots, according to Ferry, because, in an ironic reversal of Gore's argument, the ground on which the radical ecologies have taken root has already been prepared by 'romanticism, followed by fascism and Nazism' (1995: xxi). What they have started, radical environmentalists are now striving to complete. Ferry is especially critical of Deep Ecology, even though he acknowledges that its philosophy is rather complex and difficult to classify. On the one hand, he says, there is 'the love for the native soil, nostalgia for lost purity, hatred of cosmopolitanism … and the universalism of the rights of man'. On the other hand, there is 'the dream of self management, the myth of zero … growth, the fight against capitalism, racism and neocolonialism … and in favor of local power … and the right to be different'. What is one to make of these conflicting ideals? The unifying theme, Ferry reasons, is that '*the deep ecologist is guided by a hatred of modernity, by hostility toward the present*' (1995: 89). Anti-capitalist themes point to the past, to a 'world in which lost epochs and distant horizons take precedence over the present'. What we end up with, then, is a situation where the 'same obsession with putting an end to humanism is being asserted in at times schizophrenic fashion, to the point that one can say that some of deep ecology's *roots* lie in Nazism, while its *branches* extend far into the distant reaches of the cultural left' (1995: 90; my emphases). Nazi roots and radical left-wing branches. Such is the strange, hybrid, menacing, antihumanist creature that the modernist paradigm is up against and has to deal with urgently.

Concerns of this sort are exacerbated by violent acts. Take, for instance, the well-known attacks on Huntingdon Life Sciences in Britain by animal rights activists. *The Sunday Times* responded with an editorial, entitled 'Letting the mob rule', which captures quite graphically widespread middle-class fears.

> Animal rights activists *sniffed* success last week in their campaign against scientific experiments on animals. ... Like all *zealots*, they believe their methods are justified and the company's betrayal by its British financial backers is proof. Yesterday's news that American interests have saved the company signals a new phase in the battle. Anti-vivisectionists vow to *hound* the new investors once they discover their identity. ... We have a clear choice. If the law is not upheld, we become subject to the *tyranny* of cranks who oppose everything from medical research using animals to fish and chips because they fry fish. In their *nightmare* vision of society, we would not even be allowed our vegetarian diet because they would outlaw GM foods. (*The Sunday Times* 2001: 16; my emphases)

From images of 'tribal' and 'primitive' societies and the 'head-hunting' and 'cannibalistic' practices peculiar to them, to roots lying deep in European fascism and Nazism which we thought we had defeated, to visions of totalitarian tyranny reminiscent of the nightmare society of '1984' which we thought could only exist in a work of fiction. These, according to the apologists of the modernist paradigm, are the likely scenarios of what life would be like if modernity is defeated by the 'zealots of nature'. Such would be the extent of the damage that radical, and if Al Gore's ideas appear as a 'scary idealization of our past', not-so-radical environmentalists will cause if they prevail.

If there is anything, then, on which environmentalists and their critics agree this would be the magnitude of the change that is currently upon us. For both sides, it is quite clear that the modernist paradigm as we know it will not survive this change. There is no doubt, of course, that both positions are exaggerated. This, after all, is a struggle for the definition of the meaning of the world and the legitimate way of life – a struggle for both identity and power – and economy in expression or the imagination is not what one would expect to find. There is hardly an environmentalist, whether moderate or radical, who envisages a world as the apologists of modernity depict it, which is not to say, however, that what is envisaged and what could become of the vision in practice is necessarily the same thing. Nor are there many critics of environmentalism, however frightened by the more radical factions, who would deny that the world is facing environmental problems that need to be addressed in one way or another. But there is no denying either that environmentalism, even in the moderate version promoted by politicians and international bureaucracies, reflects a fundamental rupture with modernist 'physics' and 'anthropology'. If, then, the extent of this rupture cannot be doubted, the question arises as to the level at which it has occurred. Is the transformation of 'nature' into nature and of 'man' into 'human being' the result of a homologous reversal in the conditions of possibility of these visions, as one might expect? Or is there another, less perceptible and perhaps paradoxical process at work here? In the confrontation between environmentalists and their critics, such questions have not received the analytical attention they deserve. On their part, environmentalists have been meticulous in tracing the roots of the 'environmental crisis' but they have hardly ever raised the equally important question concerning the roots of environmentalism itself.[3] As for their critics, in their haste (and fright) to depict environmentalism as a cultural abomination, they tend to reduce it to primordial, reactionary sentiments, such as love of native place and soil, and a counter-modernist nostalgia for lost purity.

The environmentalists' failure to reflect on the conditions of possibility of their own vision of the world stems largely from the problematic in which they operate. In the generally acceptable and widely circulating version of the argument, the roots of environmentalism are said to lie no deeper than the 'environmental crisis' itself. The argument has been made innumerable times and probably features in every environmentalist book. Let us sketch it in its broad outline here. We have been living for centuries, the story goes, under a regime of utilitarian rapacity and greed, of improvidence and arrogance, of serious misconceptions about, if not complete ignorance of, the nature of reality and our place in it. In the process, we have managed to damage nature almost beyond repair and to put the future of humanity at risk. Recently, which is to say during the last three decades, the world has finally come to its senses. Brute reality forced itself upon us and drove home the implacable message of imminent ecological collapse. Humanity was left with a stark choice: either change its ways urgently or perish. Environmentalism was thus born.

As I have already suggested, there are serious problems with this sort of argument. To begin with, not everyone agrees on the extent of the 'environmental crisis' or, indeed, whether there is one. At stake in these disagreements are several issues: how environmental dangers are selected, how they are interpreted and presented to the public and the extent to which they can be substantiated and verified. As we have seen, critics argue that dangers are often grossly exaggerated and that many environmental facts have little or no basis in reality. Here are some specific examples from Lomborg (2001):

1. *Deforestation*: Deforestation in the Amazon was estimated at 2 per cent per year. Lomborg argues that it 'has been about 14 per cent since man arrived ... [while] at least 3 per cent of this 14 per cent has since been replaced by new forests' (2001: 114). Moreover, the idea that forests are the 'lungs of the world' is a 'myth'. While it is true that they produce oxygen, they consume precisely the same amount through decomposition when they die. 'Therefore, forests in equilibrium ... neither produce nor consume oxygen in net terms' (2001: 115).

2. *Acid rain*: Acid rain does not kill forests, as environmentalists argue. The controlled experiments carried out by the American National Acid Precipitation Assessment Program (NAPAP) demonstrated that 'even with precipitation almost ten times as acidic as the average acid rain in the eastern US (pH 4.2) the trees grew just as fast. In fact many of NAPAP's studies showed that trees exposed to moderate acid rain grew *faster*' (2001: 179).

3. *Biodiversity*: Species extinction was estimated by conservationists to be between 27000–250000 species per year. Paul Erlich, in particular, argued in 1981 that 'half of the Earth's species [will] be gone by the year 2000 and all gone by 2010–25' (2001: 249). More recent, and for Lomborg more objective estimates, including those of the United Nations *Global Biodiversity Assessment*, place the rate of extinction at 0.7 per cent per 50 years. 'An extinction rate of 0.7 per cent over the next 50 years is not trivial. ... However, it is *much smaller* than the typically advanced 10–100 per cent over the next 50 years' (2001: 255).

4. *Ozone layer*: The damage in the ozone layer, which allows more UV-B rays through the Earth's atmosphere, is not as serious as environmentalists claim. Exposure to UV-B radiation as a result of the 'hole' in the layer 'is equivalent on the mid-latitudes to moving approximately 200 km ... closer to the equator – a move smaller than that from Manchester to London, Chicago to Indianapolis, Albany to New York, Lyons to Marseilles, Trento to Florence, Stuttgart to Düsseldorf or Christchurch to Wellington'. (2001: 276)

My aim in citing these facts and figures is not to argue that they are more objective than those which environmentalists marshal in defence of their own claims. To take sides in this debate along such lines would be to reduce environmentalism to a question of scientific objectivity. My aim, rather, is to problematise the status of facts and the presumed transparency of scientific representations by noting the existence of alternative readings of the same phenomena. It is also to note that this apparent disagreement about what the same phenomena really mean points to a more general and well-known argument, namely, that facts – including, no doubt, this one – are accessed through cultural categories of perception and evaluation. This is to say, if it

needs to be said again, that phenomena become visible and acquire meaning –
relevance, significance, gravity – under determinate cultural conditions. It is to
say also that the threshold between visibility and its reverse, cultural blindness,
is not only historical, not only cross-cultural but also inter-cultural. It is to say,
finally, that visibility and value do not always coincide – what crosses the
threshold of visibility does not necessarily acquire meaning and relevance. As I
pointed out in the last chapter, the best proof of this is the 'puzzle' to which
environmentalists themselves often refer, namely, that although the general
public is well aware of environmental dangers, this awareness does not translate
into active engagement. Yet there is no puzzle here. Visibility does not guarantee
relevance or, at any rate, not of the same order across the board. What is
required, in addition, are specific cultural conditions which give rise to specific
sensibilities and sensitivities. Which brings us back to the question of the level
at which the environmentalist rupture with the modernist paradigm has
occurred and the conditions of possibility of environmentalist perception and
evaluation.

Before I turn to the critics of environmentalism and their explication of what
generates the will to save nature, it may be pertinent at this point to explore
briefly a view that goes against the current, insofar at least as it does not perceive
any conflict between environmentalism and the modernist paradigm. In this
view, environmentalism is one instance of a much wider and auspicious
phenomenon, namely, the revitalisation and further development of modernity,
which was temporarily arrested by inertia and counter-modernist forces. The
broader argument is known as the theory of 'reflexive modernisation' and is
primarily associated with the names of two sociologists, Anthony Giddens and
Ulrich Beck. I wish to explore this view here for two reasons. First, the theory
parallels in form but not in content – which is to say, it parallels superficially –
one of the main arguments of this book, namely, that at a certain fundamental
level environmentalism is a radicalised and totalised solution to an inherited and
inherent modernist problem. Discussion of this theory here, therefore, should
make clearer the differences in both perspective and level of analysis. Second, the
theory of reflexive modernisation or, at any rate, its offshoot, variously known as
'ecological enlightenment' (Beck 1992b) or 'ecological modernisation' (Mol
1996), comes up against the sort of problem that has just been discussed – the
problem, that is, which environmentalism faces whenever it explains its own
emergence as an unmediated response to the 'ecological crisis'.

As I have argued elsewhere (Argyrou 2003), the theory of reflexive
modernisation has developed in response to the poststructuralist turn in the
humanities and the social sciences and the perceived threat it posed to the
stability of the modernist paradigm. Poststructuralism's radical critique of
modernist epistemology and ethics and its refusal to be anything other than
deconstructive was understood by most apologists of the modernist paradigm as

an extreme form of relativism that left the door open to all sorts of irrationalisms, past and present. In short, as in the case of radical environmentalism, poststructuralism was perceived as a lapse into the non-modern. Some critics of poststructuralism, such as Habermas, adopted the same tactic as the critics of environmentalism themselves – they sought to demonise it. Other apologists of the modernist paradigm, like Giddens and Beck, sought to neutralise the threat by co-opting poststructuralism. The result of the second strategy is the theory of reflexive modernisation which posits as the fundamental trait of modernity not rationality or science or whatever other elements or combination of elements are conventionally used to defined the modernist paradigm. The fundamental trait, rather, is 'reflexivity' which is defined as radical questioning and doubt that includes questioning and doubting the means by which one questions and doubts. On the basis of this definition, poststructuralism is not, and cannot be, postmodern – the fear being that it could lead to the pre-modern. Its radical epistemological and ethical critique is a healthy manifestation of what it means to be a truly modern subjectivity – namely, thoroughly 'reflexive'. On the basis of this definition too, environmentalism emerges not as its critics perceive it – a cultural abomination – but as nothing less than 'ecological enlightenment'.

The argument, as it has been developed by Beck (1992a, 1992b), is that the current historical conjuncture reflects a shift from industrial society to 'risk society', which is to say, from a partially modernised society to a fully developed one. Industrial society, from the early nineteenth century onwards, was primarily concerned with the elimination of scarcity through the generation and distribution of wealth. Risk society is the society of the last few decades, which has become primarily concerned with the elimination of environmental dangers. Industrial society, according to the argument, has not been reflexive enough to foresee the hazards it would generate – nuclear, chemical, genetic. As a result, it is structurally incapable of dealing with them. There are two problems here, according to Beck. First, the nature of science is such that hazards can be studied only after the technology that generates them has been introduced into the public domain. 'Theories of nuclear reactor safety are testable after they are built [sic], not beforehand'. Second, and more importantly, decisions about the likelihood of hazards have been monopolised by experts in science and technology who inevitably hover 'blindly above the boundary of threats' (Beck 1992b: 108). Much like the type of society that preceded it, then, industrial society is characterised by a deficit in democracy. This deficit is currently being eliminated in risk society though 'the principle of division of power' and 'the liberation of doubt' (Beck 1992b: 109). Environmentalism is part and parcel of both processes. It constitutes an 'ecological enlightenment' precisely because, much like the Enlightenment at the beginning of modernity, it questions and doubts the truths handed down by the traditional authorities – 'the *scientific religion* of controlling and proclaiming truth' (Beck 1992a: 166; my emphases) – and strives for the redistribution of power in decision making.

As I have already pointed out, the major problem with the theory of reflexive modernisation – in this context, at any rate[4] – much like the problem that plagues environmentalism itself, is its disregard of culture. Reflexivity, scepticism and the perception of environmental hazards are presented as truths discovered by detached, universal subjectivities, individuals who have transcended all cultural constraints and cannot but observe reality as it really is, in its pure materiality.

> A persistent materialism and atomism underlies Beck's entire approach. His independent variables are located in the material infrastructure, and his unproblematic understanding of the perception of risk is utilitarian and objectivist. By ignoring the cultural turn in social science … Beck cuts himself off the more sophisticated and symbolically mediated discussions of risk undertaken by thinkers like Mary Douglas and Aaron Wildavsky. (Alexander 1996: 135)

If there is continuity between modernity and environmentalism, then, it is not to be located in the pure reflexivity of culturally 'uncontaminated' individual subjectivities. Beck's and Giddens' claims notwithstanding, there are no subjectivities of this sort to be found anywhere, Western societies included. I shall return to the question of continuity below, after I have examined the other case for rupture – the case put forward by the critics of environmentalism.

The attempt by some critics to reduce environmentalism to primordial sentiments of love of place and native soil, to a hatred of cosmopolitanism and modernist universalism need not retain us long. The argument is psychologistic and runs into the usual problems associated with this kind of reasoning. Because it de-culturalises and de-historicises environmentalism, it must either assume that such primordial sentiments are inherent only in some individuals, which would be arbitrary, or explain why some individuals are subject to them, while others are not – which it does not, and cannot explain without introducing social and cultural determinants. A third possibility, that some individuals are more capable than others to control such 'irrational' sentiments, begs the very same questions. Yet there may be more substance in the second strand of the modernist critique, namely, that environmentalism reflects nostalgia for lost purity, a desire for some sort of secular salvation. Although the argument is couched in rather abstract terms to have much explanatory power, it at least has the merit of pointing to some sort of social and historical considerations. The general version of this argument has been used many times before in struggles for identity and power both within the West and between it and its Others and is quite well known. As I pointed out elsewhere (Argyrou 2002), the argument 'heroises' the modernist subjectivity and stigmatises all those who allegedly lack the strength to confront reality as it is. These are the people who, as Max Weber (1946: 155) characteristically put it, 'cannot bear the fate of the times like … m[e]n' – the fate, that is, of having to live in a 'disenchanted' and therefore meaningless world. In the present context, the argument comes at different levels of analytical sophistication, and I shall turn first to the journalistic version represented by North.

> For most of its history humankind has been preoccupied with the idea of paradise. … For the western imagination, deeply imbued with the idea of mankind's fall from grace, paradise on earth or in heaven has always been a wished-for state. … Now that the idea of God, and metaphors generated by traditional religion, are less powerful explanations and templates for the way we see the world, we have needed to replace them. The natural world has replaced God. (North 1995: 192–93)

The need to believe in paradise, then, seems to be universal. Other societies still cling to their version of paradise; 'we' in the West, having killed God and 'disenchanted' the world, cannot believe in ours, which drives some people to search for a substitute. Environmentalists have found nature. The rest of us are somewhat more pragmatic. We know that 'paradise is an idea, an ideal… The paradise we dream of is not in this world, it is a dreamworld. The notion of paradise was fine in the age of religion' (North 1995: 197). Ours is no longer that age.

The second version of this argument is more academically grounded and analytically sophisticated but runs along similar lines. 'To put it simply', says Luc Ferry trying to explain what is certainly not a simple question, namely, as he puts it, the rise of 'new fundamentalisms, beginning with those that propel deep ecology' (Ferry 1995: 134) – to put it simply, then, 'in the wake of the French Revolution, we have experienced a double break with religion' (1995: 135). The first break has to do with the 'birth of secularity' or the disenchantment of the world. The second, and more recent, is the break with communism, which for the believers functioned as a religion of earthly salvation. The outcome of these two breaks is that individuals cannot find meaning anywhere else except in the private sphere.

> Emancipated from the tutelage of religious authorities, freed from dogmatic partisan lines, individuals seek the meaning of their existence *outside of religion and politics*. Meaning is now situated in the present. … We 'exist' almost like a *project* constantly setting all sorts of 'goals' [and] within these *small schemes* … our actions take on meaning. But the question of the meaning of these projects, or, if one prefers, the question of the meaning of meaning, can no longer be posed *collectively within the heart of a secular universe*. (Ferry 1995: 136)

Ferry's argument, then, is essentially Weberian. Historically, the West opted for freedom, which has the unfortunate consequence of burdening people with the responsibility of making their own lives meaningful. Having been liberated from the 'tutelage' of outside authorities, people sought meaning in their everyday pursuits but the meaning of these projects themselves remained an open question. To paraphrase Nietzsche, the wider 'why' could find no answer. Hence, some individuals are now turning to more collective, 'fundamentalist' projects which curtail freedom but provide ontological security – nationalism, for example, 'humanitarian organizations, and, last but not least … ecology' (Ferry 1995: 137).

Although not developed specifically with an eye on environmentalism, such is, in broad outline, the understanding of another apologist of the modernist paradigm – Anthony Giddens. Giddens' point of departure is modernity's emphasis on the control and subordination of nature – both internal and external nature – to culture. It is this control of nature, according to Giddens, that provides individuals in modern societies with ontological security. Being an 'internally referential system', the modernist paradigm protects individuals from coming into direct contact with the more disturbing manifestations of the natural – phenomena like madness, sickness and death, criminality, sexuality and the rawness of external nature itself. 'The ontological security which modernity has purchased, on the level of day-to-day routines, depends on an institutional exclusion of social life from fundamental existential issues which raise central moral dilemmas for human beings' (Giddens 1991: 156). Although existential questions are in this way exorcised, the system is not as foolproof as one might think (and as the modernists would have liked). To begin with, routine itself lacks 'moral meaning' and is often experienced 'as "empty" practices' (1991: 167). Second, in every individual's life there are inevitably 'fateful moments', encounters with nature in one of its disturbing manifestations. Whenever such encounters occur, 'the sense of ontological security is likely to come under immediate strain' (1991: 185). This instability and the consequent threat of meaninglessness are primarily responsible for the 'return of the repressed at the very heart of modern institutions' (1991: 202). Nature cannot be completely exorcised. Ontological security cannot be completely guaranteed. Hence, many people or, at any rate, those who 'cannot bear the fate of our times like men', seek to find security and meaning by reviving non-modern institutions or by inventing new ones. Giddens lists several such instances: the 'reconstruction of tradition', the 'resurgence of religious belief', the creation of 'new forms of religion and spirituality' and the emergence of new social movements, among them, the 'feminist movement', the 'ecological and peace movements' and 'some kinds of movements for human rights' (1991: 206–208).

For all these scholars, then, environmentalism is a movement that promises to provide people with what the modernist paradigm cannot deliver or, at least, not effectively enough, namely, existential meaning and ontological security. Yet as these authors also make clear, modernity is not about security. It is about freedom and individual autonomy. Its aim has always been to deliver individuals from the 'tutelage' of outside authorities – religion, tradition and nature itself – and to restore them to where they truly belong – themselves. As Hegel pointed out reflecting on the modernist paradigm, such has been the trajectory of the entire human history, the monumental battle of 'Spirit' over 'Matter' which had been won only in the Europe of Hegel's time and whose benefits – autonomy and freedom – have been enjoyed in Europe ever since. 'Matter has its essence out of itself; Spirit is *self-contained existence*. Now this is Freedom, exactly. For if I am dependent, my being is referred to something else which I am not. ... I am free, on the contrary, when my existence depends upon myself' (Hegel 1991 [1894]: 17). Such was also Kant's view. Debating the nature of the modernist paradigm before Hegel, Kant made clear that '*Enlightenment is man's emergence from his self-incurred immaturity*'. Immaturity, Kant says, 'is the inability to use one's own understanding without the guidance of another [person]. ... The motto of enlightenment is therefore: *Sapere aude*! Have courage to use your *own* understanding!' (1970a [1784]: 54). It does take courage to use one's own understanding and to think without the guidance of 'another'. Autonomy comes at a high price. Severing the bonds of dependence also means relinquishing the protection and security they provide. Yet it is a price worth paying. For Kant, as much as for every modernist that came after him, freedom is what makes us truly human. Without it, we would be leading the life of 'domesticated animals', of those 'docile creatures' that do not 'dare take a single step without the leading-strings to which they are tied' (Kant 1970a: 54).

Recycling the same ideas two centuries later, Giddens argues that dependence of this sort is a pathological condition. Some individuals find 'that the freedom to choose is a burden and they seek solace in ... overarching systems of authority'. Yet 'a predilection for *dogmatic authoritarianism* is [a] pathological tendency': the individual 'identifies with a dominant authority on the basis of projection' (Giddens 1991: 196). It is no wonder, therefore, that to those who wish to lead such a life, the modernist says emphatically:

> To the person who cannot bear the fate of the times like a man, one must say: may he rather return silently, without the usual publicity built-up of renegades, but simply and plainly. The arms of the old churches are opened widely and compassionately for him. After all they do not make it hard for him. One way or another he has to bring his 'intellectual sacrifice' – that is inevitable. If he can really do it, we shall not rebuke him. (Weber 1946: 155)

We shall not rebuke 'him' but neither are we prepared to allow 'him' to take the rest of us down the same road. As we have seen, the critics of environmentalism have made this abundantly clear.

It should come as no surprise that environmentalists agree with their critics that the modernist paradigm lacks meaning and that life in it is often experienced as a string of 'empty practices'. 'Coming of age in the modern era marks a passage into emptiness', complains Charlene Spretnak in 'The Spiritual Dimension of Green Politics'. Having tried the 'endlessly varied and attractively marketed ... diversions' of modernity, people come to realise that 'there is no inner life in a modern, technological society' (Spretnak 1984: 230). Al Gore (1992: 231) concurs and notes an unfortunate confusion. 'Many who feel their lives have no meaning and feel an inexplicable emptiness and alienation simply assume that they themselves are to blame'. Yet the blame should be placed elsewhere. 'Ironically, it is our separation from the physical world that creates much of this pain'. Nor should it come as a surprise that environmentalists recognise, indeed, emphasise as much as their critics that the quest to save nature is also, inextricably, a religious quest. As we have seen in the last chapter, for many environmentalists, particularly of the more radical persuasion, nature is not merely fragile but also sacred, pollution not merely material but also symbolic contamination – defilement and sacrilege. 'The pain we feel over the environmental crisis is *not* solely a self-interest ... to retain some wilderness in which to hike', says Gottlieb (1996a: 11). 'Our response is ... a *spiritual* one; ... it involves our deepest concerns about what is truly of lasting importance in our lives'. Hence, the destruction of nature leaves 'a deep sense of desolation', the feeling 'that an enormous and unrectifiable sacrilege has been committed'. Al Gore (1992: 263) concurs once again and calls for an 'environmentalism of the spirit'. Armed with faith, he points out, 'we might find it possible to resanctify the earth, identify it as God's creation, and accept our responsibility to protect and defend it'.

We should not find the agreement between environmentalists and their critics about the existential shortcomings of the modernist paradigm and the deeper significance of environmentalism surprising because both groups share the same conviction: that the change currently upon us constitutes a radical rupture with the modernist paradigm. To paraphrase Bourdieu, this conviction is the complicity that unites the two groups in hostility. Having discussed the hostilities at some length, it is now time to turn to the complicity itself. It is time to take a first step beyond the phenomenology of change and examine whether environmentalism is indeed such a rupture.

'The Age of the World Picture'

To unpack the silent complicity between environmentalists and their critics, two issues need to be addressed. The first is whether we should accept at face value the modernist claims about modernity: that it is a disenchanted universe suitable only for those who are prepared to live like 'men' – free and with no metaphysical safety lines attached. The second and related issue is whether we should accept at face value the modernist claims about environmentalism: that it provides an escape route to those who cannot bear the 'fate of times' and constitutes a secular religion of one sort or another. Before any of these questions can be answered, however, it is necessary to raise an even more fundamental one, namely, the question concerning the conditions of possibility of both the modernist and environmentalist visions of the world.

As we have seen, modernity is said to be an 'internally referential system', a system, that is, whose forms of knowledge, and no doubt power also, are ultimately legitimised with reference to themselves. Beyond the obvious problem of circularity, the question arises as to how it can be imagined in this way. I am not raising this as a specifically historical question, although it can be, and has, in fact, been raised in such terms many times. My interest, rather, lies primarily in the logical and epistemological operations necessary for the construction of such a system. It should be clear that, by definition, an internally referential system contains all the referents peculiar to it. This is to say, among other things, that it also contains the referent that does the referring. What sort of operation is involved in this deceptively simple act of imagining the modernist paradigm as an 'internally referential system'? Where does one need to go to have a view of it as such a system – a closed, self-contained, circumscribed totality? Apparently, this can only be an external position. But where exactly is this strange position to be located? What prompts anyone to take up such a position, and what *exactly* happens when one arrives there?

These sorts of questions have not always received the analytical attention they merit either by those who are critical of the entire exercise or by those who must assume such a position. For the former, the possibility of stepping outside the system, which is to say in effect, outside society and history, is an illusion, a myth, nothing more than a metaphysical fancy. Whatever happens when one makes such a metaphysical leap and lands on the other side of reality is of little relevance. For the latter, both the position and the positioning are given a priori; and because they are given in this way, there is nothing more to say about them by definition. Yet if, as Heidegger points out, metaphysics grounds an age, and if, as both environmentalists and their critics claim, we are currently witnessing the passing of an age and the emergence of a new one, the metaphysics of both the modernist and environmentalist paradigms is a domain of inquiry that we cannot afford to overlook.

The second and related issue that must be addressed is whether, in what way and to what extent the quest to save nature is also a quest to save individual subjectivities from the presumed meaninglessness of the modern condition. As I have just suggested and will explore in detail below, the question as to whether the modernist paradigm is intrinsically meaningless remains open. But neither can we take for granted the claim that environmentalism provides an answer to the 'problem of meaning'. To do so would be tantamount to de-historicising and de-culturalising the issue as, indeed, both environmentalists and their critics have done. There is no such thing as the 'problem of meaning' in the abstract, outside of determinate historical and cultural conditions. Even death itself cannot be considered intrinsically meaningless. As Max Weber (1946: 356) points out, death becomes 'meaningless precisely when viewed from the inner-worldly standpoint', the point of view, that is, of a here and now that cuts itself off from the possibility of a beyond. The peasant, Weber says, 'could die "satiated with life". The feudal landlord and the warrior hero could do likewise. For both fulfilled a cycle of their existence beyond which they did not reach'. A satiated life, of course, does not necessarily make death welcome. But unlike an incomplete and unfulfilled one, neither does it make it necessarily meaningless.[5]

The problem of meaning arises under specific social and cultural conditions and if environmentalists have come up against it, we need to know what these conditions are. To turn to Max Weber (1946: 280) once again, although, he says, the idea of salvation is very old, '[it] attained a specific significance only where it expressed a systematic and rationalized "image of the world" and represented a stand in the face of the world... . "From what" and "for what" one wished to be redeemed ... depended upon one's image of the world'. Weber takes this in a stride but if we are to address the sort of questions that he does not, one or two pauses are necessary. In Weber's schema there is first, a 'rationalised' image of the world; second, there is what one experiences in the 'face of the world' – for Weber not so much an image as the 'actual' world itself; and third, a conflict between the two. Having taken a stand in the world, people come to see that its 'face' does not fit their 'rationalised' image of it. When this happens, Weber goes on to say, the actual world, or 'something in the actual world ... is experienced as specifically "senseless"' (1946: 281). The question that Weber does not explore concerns the conditions of possibility of the 'rationalised' image of the world. As far as he was concerned, such an image was the outcome of an increased intellectualisation of life. Yet this hardly explains the logical and epistemological operations involved in the construction of the image nor does it address the issue of its ontological necessity. To return to environmentalism, we know very well what the 'rationalised' image of the world is in this case. It is an image in which humanity and nature are no longer separate or, in the more radical view, are essentially and fundamentally the same. As we have seen, 'man' has been cut down to the size of the 'human being' to fit in nature; and 'nature' – the intractable domain of utility and danger – has been transformed into 'Mother Earth' to accommodate the human being as a giving and caring mother. We also know what it is in the actual world that, on the basis of this image, is experienced as 'specifically senseless'. It is the destruction of nature by 'man'. What we do not know is how exactly the 'rationalised' environmentalist image of sameness has been constructed. Without an adequate understanding of the process that produced this image it would be simplistic to assert that the quest to save nature is also a quest for meaning. It would be equally simplistic to say that this will to meaning is unique to environmentalism. I shall examine the case of environmentalism in detail in the next chapter. For the time being, I wish to return to the other 'rationalised' image of the world – the image of the modernist paradigm as an 'internally referential system' – and the set of questions raised already: What are the conditions of possibility of this image? Where does one need to go to have such a view of the world? And what happens when one arrives there?

In trying to answer these questions, I shall use as a starting point an important essay by Heidegger entitled 'The Age of the World Picture'. The essay is concerned with the metaphysics of the modernist paradigm and examines in a systematic fashion the logic that produces an image of modernity as a totality and a system. A brief discussion of this essay here, therefore, should prepare the ground for addressing the set of questions raised above.

Let us first say that for Heidegger, the metaphysics of the modernist paradigm is different from those of earlier periods in the history of European thought in that it conceives of the world as a 'picture'. A picture of the world is an image of the world in its totality, a universalising vision, a worldview in the literal sense of the term. Yet not all worldviews are 'world pictures'.

> World picture, when understood essentially, does not mean a picture of the world but the world conceived and grasped as a picture. What is, in its entirety, is now taken in such a way that it first is in being and only is in being to the extent that it is set up by man, who represents and sets forth. Whenever we have the world picture, an essential decision takes place regarding what is in its entirety. The Being of whatever is, is sought and found in the representedness of the latter. (Heidegger 1977a: 129–30)

There is, then, a difference – and a fundamental one at that – between a picture of the world, such as those of other ages and other cultures, and 'the world picture' unique to the modernist paradigm. The difference has to do with making a decision about 'what is, in its entirety', what exists and could possibly exist. The critical point for Heidegger, however, is not so much the decision itself as who makes it, which is to say in effect, who assumes responsibility for the existence of the world. In the modern era, it is 'man' who makes this decision; it is he who decides what is 'in being' and can be 'in being', he who represents and sets 'forth'. In making this decision, 'man' creates the world. He becomes a subject and the world his object or, to stay with Heidegger's metaphor, the painter and the world his picture. Such was not the case in the metaphysics of the Middle Ages, according to Heidegger (1977a: 130): 'That which is, is the *ens creatum*, that which is created by the personal Creator-God as the highest cause. Here, to be in being means to belong within the specific rank of the order of what has been created – a rank appointed from the beginning'. Nor was 'man' the creator of the world during 'the great age of the Greeks'. On the contrary, for the ancient Greeks:

> That which is does not come into being at all through the fact that man first looks upon it, in the sense of a representing that has the character of subjective perception. Rather man is the one who is looked upon by that which is; he is the one who is ... gathered towards presencing, by that which opens itself. (1977a: 130–31)

For Heidegger, then, the Greek understanding of the world is the reverse of modern 'man's' understanding. The world exists not because it is represented by 'man' but because it itself appears to him in the guise of the different beings that constitute the world and invites him to 'apprehend' it. The world lies in waiting, as it were or, better still, in flux and calls on 'man' to catch it as it rushes by. If 'man' does apprehend it, the beings of the world emerge into the open. If he does not, they pass him by, withdraw and conceal themselves. Heidegger uses the Greek term *aletheia* to convey this interplay between emergence and concealment. The term means 'truth' but its real significance lies in its etymology. It is a conjunction of the negating '*a*' and the world '*lethe*', which is the state of being forgotten. Literally, then, truth as *aletheia* is the negation of this state of being. It is the 'remembering' of things through apprehension – not their correct representation, as the modernist epistemology would have it. The critical point for Heidegger is the independence of the beings of the world from 'man'. If 'man' fails to apprehend them, they do not thereby cease to exist. Rather, he misses the chance to reveal or 'disclose' what is already there.

To clarify further both the distinction between 'apprehension' and 'representation' and the distinction between a picture of the world and 'the world picture', it may useful at this point to turn to Kant, whose work Heidegger has very much in mind when he himself makes these distinctions. As is well known, Kant conceived of his *Critique of Pure Reason* as a Copernican revolution. Copernicus explained the apparent movement of the stars as being partly the result of the movement of the observer; Kant explained the nature of Being by suggesting that beings themselves conform to the human mind. 'A new light must have flashed on the mind of the first man (*Thales*, or whatever may have been his name) who demonstrated the properties of the *isosceles* triangle', says Kant (1934 [1781]: 10) in the Preface of the *Critique of Pure Reason*. 'For he found that it was not sufficient to meditate on the figure, as it lay before his eyes' in order to understand its properties. He found 'that it was necessary to produce these properties' himself and impose them on the triangle. Before the new light flashed on his mind, Thales (or whatever may have been his name) behaved very much like the Greek 'man' that Heidegger has in mind. He was 'looked upon' by the isosceles triangle and was 'summoned' to apprehend it and bring it out of its concealment. Once the new light flashed, Thales created the isosceles triangle as we now know it by deciding what its properties were, that is, by representing it as he himself saw fit. Before the light, the triangle was what it was in its own being; after the light, its being became what Thales said it was. Thales and the men who followed his lead at the end of the Middle Ages, such as Bacon and Galileo, Kant (1934 [1781]: 10) goes on to say, 'learned that reason only perceives that which it produces after its own design'. Having acquired this knowledge, they realised also that they 'must not content to follow … in the leading-strings of nature but must … compel nature to reply to [their] questions'.

What Thales did for the isosceles triangle, Kant did for the world writ large. If reason perceives only what it produces after its own design, it was imperative to decide what this 'design' was. On the basis of this decision it would then be possible to decide what for reason is real and what not. The 'design' was, of course, the 'pure intuitions' of time and space. Within these boundaries lies the empirical, beyond them the metaphysical – a realm into which people can no doubt venture and speculate about but are warned by Kant not to because it always leads to contradiction. This, in effect, is the 'essential decision' taken regarding what exists in its 'entirety' which Heidegger laments as a disastrous event in the history of Western 'man' and of Being itself. This is the decision that marks the transformation of the world into a 'world picture' – a human creation – and the onset of the modern age. This is also what it ultimately means to say that the modernist paradigm is an 'internally referential system'. It has to be. On the basis of this decision, nothing 'real' can possibly exist outside the system, nothing to guarantee and hence legitimise the inside. From now on, what lies inside has to guarantee and legitimise itself by itself.

What prompted Kant to make this 'essential' decision about what exists in its 'entirety'? In the usual version of events, Kant was concerned with finding a secure basis for human knowledge. He was reacting, it is often said, to Hume's unsettling revelation that causation was not an inherent part of the empirical world but a habit of the human mind. He himself points out that, in addition, he wished to undercut the monopoly of truth exercised by the Church. Heidegger (1977a: 90) provides a broader and culturally more substantial account. 'At the beginning of the modern age', he says, 'the question was freshly raised as to how man … can become certain and remain certain of his own sure continuance, i.e., his salvation'. This was the time, Heidegger (1977a: 148) goes on to say, when 'man' was beginning to liberate himself 'from obligation to Christian revelational truth and Church doctrine'. Yet this liberation of 'man', whether he understood it or not,

> is always still freeing itself from being bound by the revelational truth in which the salvation of man's soul is made certain and is guaranteed for him. Hence liberation *from* the revelational certainty of salvation had to be intrinsically a freeing *to* a certainty in which man makes secure for himself the true as the known of his own knowing. That was possible only through self-liberating man's guaranteeing for himself the certainty of the knowable. Such a thing could happen, however, only insofar as man decided, by himself and for himself, what, for him, should be 'knowable' and what knowing and the making secure of the known, i.e., certainty, should mean. (1977a: 148)

One of the first men, according to Heidegger, to secure for himself 'the true as the known of his own knowing' was Descartes who found this sort of certainty in the thinking self – *cogito, ergo sum*. One of the men who 'decided, by himself and for himself, what, for him [and for all other men], should be knowable' was Kant.

Both had to make a decision about what 'should be knowable' and hence 'real' because there was no longer anyone else to make it for them. The revelational truth of the Christian Church was by now collapsing and with it, the structure of the world itself. Under the circumstances, 'man' had no option but to step in and assume sole responsibility for the world and for himself – no option, that is, if he wished to become once again certain and remain certain of 'his own sure continuance' – his salvation.

The attempt to find a secure basis for knowledge at the dawn of modernity, then, was not an intellectual exercise for the curious and the idle, a concern cut off from the cultural conditions of the time. It was imposed by those conditions and was guided by a logic peculiar to them – an objectifying, totalising, universalising logic. Having rejected the picture of the world that was revealed to him and guaranteed by the Christian God, 'man' assumed himself the role of the creator and guarantor of the world. Having assumed this role and become the Universal himself, he had no option but to *universalise*, which is to say, as a bare minimum and as the condition of possibility of everything else he would be doing, no option but to decide the parameters of the possible, to demarcate the boundaries of the real, to construct the structure of the Whole, to draw a firm outline of the image of the world and of himself. In short, 'man' had no option but to produce an entire and entirely new ontology. Which brings me to the topography of this grandiose construct and the two questions raised above: Where does 'man' need to go to accomplish such an extraordinary feat? And what happens exactly when he arrives there?

In the quest to identify the *topos* which 'man' must occupy in order to construct a totalised image of the world, it may be instructive to return briefly to environmentalism and one of its chief claims. As we have seen, environmentalists are in the habit of pointing out that we are the first generation to see our planet from space and that looking from such a distance the world appears very different from the way in which earlier generations perceived it on the ground. From this position outside the planet, we can clearly see that nature is fragile – indeed, both fragile and beautiful. We can see also that nature is part of us as we are part of nature – which presumably makes 'man's' view of nature short-sighted not only metaphorically but also literally. What is one to make of these claims? Let me first say that I have chosen to discuss them in relation to the topography of the universal because they are pregnant with important implications, even though the ones that I shall draw are quite different from those pursued by environmentalists. There is no doubt, of course, that we have seen our planet from space and that the place or places from which the photographs were taken are as real as anything else we know. This, however, does not automatically make the environmentalist claims literally true. As I have already pointed out, looking at the earth from space with no prior knowledge, there is hardly any way to deduce from what one sees that nature is 'fragile'. Nor can we literally see that we are part of nature, much less that nature is part of us, if not for any other reason, because of the simple and obvious fact that we do not actually see any human beings. From such a distance, another observer would not even know that the planet is inhabited.

What we see from space is a bright and colourful planet, which we recognise as the Earth, suspended in the void and surrounded by utter blackness. If we 'see' more than this, it is only because this particular image acquires specific meaning and value in a wider cultural context – the context of our 'rationalised' image of the world. Let us note some of the underlying assumptions: the sheer vastness, emptiness and inhospitability of space, the staggering age of the universe, its incomprehensible size, the unimaginable power of the forces it contains, its unknown fate, its total indifference to human affairs, the unfathomable reason for its existence. And side by side and in relation to these, the certainty of our finitude and the unfathomable reason for our own existence. In a cultural context such as this – a context of precarious existence and profound cosmic loneliness – Earth does appear fragile and we as little creatures embedded in nature and hopelessly dependent on it. Yet this vision is not so much an actual perception as a meditative reflection, not so much a literal way of speaking about the nature of nature as a metaphorical way of expressing concern about what we think is the nature of the human condition. This is to say, among other things, that the condition of possibility of the vision of the world in its entirety is not literal externality, a real position in space, but a symbolic one, a position in the imagination or, if one prefers, a metaphysical position.

One of the few anthropologists concerned with the environment to have dealt with the question of literal and symbolic externality, even if not in quite the same terms, is Tim Ingold.[6] In an insightful paper, Ingold reflects on the frequently used expression 'global environmental change' and makes the point that it is a rather paradoxical statement. To speak of the environment is to speak of what surrounds us. To speak of the Earth as a globe is to assume a position of externality, such as the position of schoolchildren who view the model globe in the classroom. It is to say, in effect, that it is we who surround the environment, not the other way round. The problematic nature of this expression, Ingold goes on to say, becomes even more apparent when we consider that in other societies, including medieval Europe, the world was perceived not as a globe but a sphere. Unlike a globe, a sphere is something that surrounds us, a true environment (see Figure 1). How, then, is one to explain the inconsistency in the expression 'the global environment'?

Ingold's argument is that the image of the world as a globe encapsulates the nature of Western ontology and the way in which people relate to the world. Instead of direct engagement with the world, which produces an understanding of it in particular contexts, people detach themselves from the world, as though they were free-floating, context-free subjectivities, and reconstruct it in the imagination as a totality. In the modernist ontology of the West, Ingold points out, meaning does not lie in the relational context of the perceiver's involvement in the world, but is rather inscribed upon the outer surface of the world by the mind of the perceiver. To know the world, then, is not a matter of sensory attunement but of cognitive reconstruction. And such knowledge is acquired not by engaging directly, in a practical way, with the objects of one's surroundings, but rather by learning to represent them, in the mind, in the form of a *map*. (2000: 213)

There are several problems with the way in which Ingold's argument is couched, and I shall return to them in the next chapter. Here it should suffice to point out that there can be no such thing as 'sensory attunement' to the world which is not also *always already* a cognitive reconstruction of one sort of another. Ingold needs to make a radical distinction between different ways of knowing the world because, in the environmentalist vein, his aim is to show that the modernist ontology is responsible for the destruction of nature. This problematic aspect of the argument aside, there is an important point to remember from Ingold's analysis, namely, that the representation of the world in its totality – 'the world picture' – does not require literal perception. One does not need to physically step outside the world to see it as a whole, like an astronaut. One can do so more easily in symbolic terms, by taking up a position in the imagination. What is required is the sort of subjectivity that transcends particular contexts to such an extent so that the world itself ceases to be *a* context and comes to be seen *in* context, as part of a wider reality that contains it. What is required is a transcendental subjectivity that sees the world from the point of view of the cosmos itself.

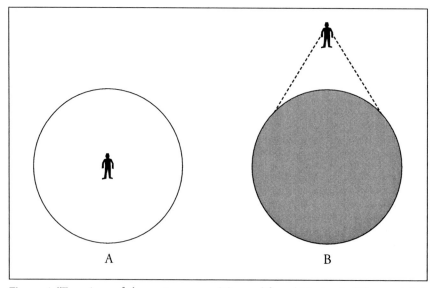

Figure 1 'Two views of the environment: (A) as a lifeworld [sphere]; (B) as a globe'. Reproduced from Ingold, T., *The Perception of the Environment*, Routledge, 2000, p. 209.

As Heidegger suggests, a leap beyond the world is taken every time a decision is made about what exists in its entirety. It must be taken even when one's aim is precisely to argue against transgressing the boundaries of the real, since, as Kant pointed out, it inevitably leads to contradiction. For all his warnings, Kant could not avoid contradiction himself. He could not because in order to set time and space as the boundaries of the world, he had to transgress them himself at least once. As is well known, for Kant, how things are in themselves, that is, independently of the pure intuitions of time and space, which are given to us a priori, we cannot know. What we *can* know is things as phenomena, as they appear in relation to the human mind. Poststructuralists of the likes of Foucault have discussed the contradiction entailed in this argument extensively. For the purposes of the present discussion, however, I shall turn to a less politically charged commentary by Collingwood in *The Idea of Nature*.

There is something highly problematical with Kant's notion of the thing in itself, Collingwood points out. It is impossible to state it 'without flatly contradicting yourself' (1945: 116). Kant's argument 'implies that there must be minds, and must be things in themselves; if these do not exist the whole argument falls to the ground'. Minds must exist because they produce the pure intuitions of time and space; things in themselves must also exist because they produce the phenomena that minds perceive and understand. Yet, Collingwood goes on to say, 'since we can know only phenomena, we cannot on the argument know either minds or things in themselves. If so, how can we say that they exist?' (1945: 117). The argument, then, 'flatly' contradicts itself because it posits and hence claims to know the means through which one can know, which on the basis of this positing are unknowable. It contradicts itself as soon as it is uttered because Kant had to step outside the world of phenomena to draw the boundaries that enclose it, only to return and claim that there is no 'outside'. The argument is contradictory, in short, because Kant burdens himself with an impossible double task, which is the hallmark of the modernist ontology: he must play both the role of the creator of the world and of a creature in it.

The discussion so far has more or less reiterated what in recent decades has acquired a sharper political edge through the poststructuralist efforts to de-centre 'man': To decide what exists in its entirety, the modernist subjectivity must venture beyond the world – the impossibility of which is manifested in contradiction. It is now necessary to unpack this generalisation and ask what happens exactly when this metaphysical leap is taken. As I have already suggested, questions of this sort are not raised very often in the social sciences largely because of scientific pretensions. No one can step outside society and history and any such claim can be nothing more than metaphysical fantasy, a fiction, an illusion. That may be the case. Nonetheless, people do venture beyond the world, and if metaphysics is ultimately responsible for the emergence of an era, as Weber has argued about capitalism, if it grounds an age, as Heidegger suggests, we cannot afford to overlook the fictions that people produce. Anthropologists, at any rate, have always insisted on treating other peoples' mythical realities seriously. Not to do so with their own would make their pronouncements about other cultures not only trivial but also condescending and patronising. Let us, then, turn to this task with the appropriate anthropological seriousness.

In the modern age, it is often said, 'man' has taken the position of God. The statement is usually meant to underscore 'man's' arrogance or, as in poststructuralist discourses, to show that this impossible feat is responsible for all sorts of contradictions in the claims of the 'human sciences'. The argument is not inaccurate but further reflection suggests that the equalisation of 'man' with God is highly problematical. I should hasten to add that I am not raising a theological issue here – my point is purely epistemological. There is no doubt that 'man' has displaced God, at least insofar as he has become the only relevant Subject of the world. Nonetheless, there is a fundamental difference between their respective visions. God's vision of his creation is 'panoptic'. This is to say he sees everything, both the Whole as a whole *and* all its constituent parts in their uniqueness and individuality down to the most minute and insignificant being. This, at any rate, is what the Christian fathers have long argued. 'Man's' vision of his own creation – his 'world picture' – on the other hand, is not of this sort. The distance that 'man' places between himself and the world in order to take it in as a whole, which is to say, the level of abstraction involved in this sort of conceptualisation is such that 'man' loses sight of individual beings and their particularities. The greater the distance, the more individual things become blurred, and the more the world appears as a unity. The limiting case of this process, epitomised by Hegel, is that 'man' places such an enormous distance between himself and the world – or reflects at such a high level of abstraction – that he reaches the point where he sees nothing. This is not to say 'nothingness'. It is to say, rather, 'no-thing': no individual things, no particular beings but a seamless, indistinguishable whole – Being pure and simple.

As Collingwood points out, Hegel rejected Kant's claim that the thing in itself is unknowable. He argued, on the contrary, that it is 'the easiest of all things to know'.

> [The thing in itself] is simply pure being, being as such, without any particular determinations whether qualitative or quantitative, spatial or temporal, material or spiritual. The only reason why it seems to be unknowable is because there is nothing particular in it to know; it has no characteristics to distinguish it from anything else, and so when we try to describe it we fail, not because we cannot understand the mystery of its nature but because we understand perfectly well that there is nothing there to describe. Being in general is nothing in particular; so the concept of pure being passes over, as Hegel puts it, into the concept of nothing. (Collingwood 1945: 121)

Such, then, is the ultimate vision of the world – a vision in which the world of common sense and everyday experience disappears completely. It is as if the beings of the world melt away and fuse with one another so that it is no longer possible to make any distinctions whatsoever, as if the space between the beings of the world that keeps them apart is suddenly drained off so that every-thing collapses onto every-thing else to form an immense, completely smooth, completely amorphous whole – Pure Being. To be sure, not very many people ever come to conceptualise the world in such terms. Nonetheless, most are familiar with the process that ultimately leads to this strange, metaphysical vision. Certainly, everyone is familiar with the experience of losing sight of detail as a result of spatial separation. It happens when the distance between the observer and the object of observation is too great – one sees the forest, people say, but no longer the trees. But equally importantly, it happens also when the distance is too small, in which case what one sees becomes blurred. I shall examine the latter case in detail in the next chapter. Here it should suffice to point out that no distance produces exactly the same result and infinite distance – a vision of Pure Being.

Although I have been using space as a metaphor, many people are also familiar with the actual process at work, namely, in the case of distancing, the process of conceptual abstraction and generalisation. To stay with the tree example, here too, people lose sight of them, of their sensible properties and distinctive features to begin with, and as one moves up the ladder of abstraction and generalisation to species, genus, family, order, class, phylum and so on, of the trees themselves. Here too, what becomes apparent is not what is different about the beings of the world but what is the same, not what divides and keeps them apart but what brings them together in increasingly wider totalisations. It is in this too that we must locate the difference between God's vision of the world and the vision of the deified Western 'man'. The former is '*pan*-optic' – it sees all, both the whole as a whole and all its individual parts in their individuality and uniqueness. The latter would be more accurately described as '*syn*-optic' – it sees all, but everything put together, the world as a unity. God has this ability because he sees things directly – he creates them through the sheer act of thinking them. The deified modernist subjectivity, by contrast, has no such ability. Its thinking is limited because it is mediated. It sees things through things; it creates them, which is to say, invests them with meaning, in relation to other things. Its vision, in short, is synoptic precisely because it is relational.

The primary effect, then, of the metaphysical move by which 'man' steps outside the world to take it in as a whole is the levelling of difference and the equalisation of the things of the world. The further beyond the world he moves the greater the equalisation and the more intense the revelation of sameness, the limiting case being a sameness that rests on complete identity of everything so that 'no-thing' is any longer visible – Hegel's Pure Being. There are countless examples of this process in the literature but here I shall stay with the historical examples that contrast the modernist ontology with that of the ancient Greeks. Collingwood, for example, points out that 'for Aristotle, nature is made up of substances *differing* in quality and acting *heterogeneously*: earth *naturally* moves towards the centre, fire away from the centre, and so forth'. It seems that Aristotle and his contemporaries were very much within the boundaries of the world – or were 'looked upon' by things, as Heidegger would have it. If they could discern natural differences in the quality of things, it is only because they maintained what the phenomenologists call a 'natural attitude' towards the world, a common sense perspective. By the time of Bacon and Galileo, Collingwood goes on to say, these differences are stripped away. 'For the new cosmology there can be no *natural* differences of quality; there can only be *one* substance, qualitatively *uniform* throughout the world, and its only differences are therefore differences in quantity and of geometrical structure' (1945: 98; my emphases). Heidegger makes a similar point. In contrast to the Aristotelian doctrine of differences in quality, the modern ontology posits that 'no motion or direction of motion is superior to any other. Every place is *equal* to every other. No point in time has preferences over any other' (1977a: 119; my emphasis).

No special times, then, and no unique places. Such things would still be possible but only in the common sense world of everyday life, only from the perspective of those firmly anchored to the ground. Looking at the world from the outside, time and space become uniform or, as it is sometimes pointed out, 'empty'. They do so because they are divested of the particularities – memories, experiences, feelings – which the 'natural attitude' attaches to them. The same happens to everything contained within these boundaries. From now on, nothing – no-thing and no being – in the sciences of nature would occupy a privileged position, nothing would have preference over anything else. Privilege and preference are attitudes of the everyday not found in the modernist ontology on which the sciences of nature are based. The new ontology reduces the beings of the world to a common denominator, a universal feature, the same substance, and the only way to tell them apart is in terms of how much of this same substance each contains. As for the nature of the substance itself, its precise determination depends on how far outside the world one positions oneself and, as the discussion in the next chapter will make clearer, from which end of the temporal spectrum one is looking – whether, that is, one is looking from the origin of time or its end. At a certain distance from the origin, for instance, the universal feature becomes proteins and nucleic acids, which means that human beings become essentially and fundamentally the same with all sorts of non-human beings – until recently a startling revelation in the context of everyday life but quite ordinary since the rise of environmentalism. 'An oak tree and I are made of the same stuff', points out a popularizer of science (Sagan 1980: 33; see Figure 2). No doubt, he goes on to say, 'we humans look rather different than a tree. Without a doubt we perceive the world differently than a tree does'. But these differences are relevant and meaningful only in the context of everyday life. 'Down deep, at the molecular level, trees and we are *essentially identical*' (1980: 38; my emphases). Down deep at the molecular level is also deep down in the well of time, far away in the past towards the origin of life, as evolutionary biology teaches.

Figure 2 'Close relations: an Oak tree and a human'. Reproduced from Sagan, C., *Cosmos*, Macdonald Futura Publishers, 1980, p. 33.

I shall return to this issue in the next chapter, since the underlying metaphysics constitutes one of the directions that environmentalism takes to produce its central argument, namely, that humanity and nature are essentially and fundamentally the same. For the moment, it may be useful to summarise the foregoing discussion. It would seem that inherent in the modernist paradigm is a tendency for metaphysical totalisations of an *epistemic* nature – a tendency imposed by the need to make a decision about what exists in its entirety. The modernist subjectivity ventures beyond the world because it is only from such an external position that the boundaries of the world can be drawn and knowledge of what exists guaranteed. Once 'there', it visualises the world synoptically, as a unity of different beings with the same substance. The effect of this metaphysical move, in other words – which, as we shall see, parallels the effect of another move in the opposite direction – is the levelling of difference apparent in the common sense world of everyday life.

Yet 'man' is not merely the Subject of the world. He is also a being in the world so that among the world's other objects he also encounters himself. Hence, the modernist ontology has produced not only the sciences concerned with the natural world but also sciences devoted to this special object – the 'moral sciences', as the philosophers of the Enlightenment called them. This is to say that from his cosmic perspective, it is not only the things of the world that appear to him the same but also 'men' themselves – for a long time only certain men but more recently, of course, all men, including those 'other men', namely, wo-men. It is to say that the most fundamental effect of 'man's' metaphysical venture in relation to the *human world* is the levelling of social difference and the equalisation of people. It is to say, finally, that inherent in the modernist paradigm is a tendency not only for *epistemic* but also *ethical* totalisations and abstractions.

Pure Humanity

The central question raised in this chapter was whether we should accept the claim that environmentalism constitutes a radical rupture with modernity – a claim that, as we have seen, is the complicity that unites environmentalists and their critics in hostility. I suggested that at the phenomenological level, rupture is undoubtedly the case, but at a deeper and more profound level, this rupture is made possible, paradoxically, by a fundamental continuity between the two. I suggested also that if there is anything radical about environmentalism, this is nothing other than the radicalisation of the modernist logic. This claim cannot be fully substantiated until the next chapter where I shall examine in detail the conditions of possibility of the environmentalist vision of the world and the key argument that humanity and nature are essentially and fundamentally the same. In the meantime, and as a first step towards this goal, it has been necessary to establish the nature of the modernist logic itself – the logic of the Same – by following its deployment in one of the two possible directions that lead to the Same: towards distancing and abstraction as opposed to proximity and contraction. In the preceding section, I explored in broad outline the 'physics' of this logic, the sort of non-human reality it has produced. In this section, I turn to its 'anthropology' and the modernist vision of social reality. My aim, in the first place, is to question one of the most persistent and enduring modernist claims – the 'disenchantment of the world' – a claim all too often taken for granted even by critics of the modernist paradigm and one of the weapons used by modernists themselves to criticise and dismiss a whole host of ideas and practices, including environmentalism.

As I have already suggested, and pursued elsewhere in greater detail (Argyrou 2002), there is a long genealogy of thinkers – from Kant to Max Weber to Giddens and Habermas, to mention only a few names – who depict the modern condition as both a blessing and a predicament – the latter condition in a strategic and calculated manner, since the greater the predicament, the higher the profits of recognition for those who bear it. It is only in modernity, according to the argument, that human beings become what they ought to be – truly free, autonomous individuals. And it is under modern conditions that individuals, having liberated themselves from the tutelage of higher authorities, come to realise that there are no ontological guarantees in life. Corollary to this is the argument, produced for both internal and external consumption, that some individuals, those who cannot bear the 'fate of the times', turn to old religions and metaphysical systems or invent new ones in search of ontological security (or adhere to them stubbornly for this very reason). As we have seen, critics of environmentalism explain it, or to be more precise, explain it away (incidentally, in a similar vein as critics explain nationalism away) in these very terms: as a rediscovery of a pre-modern metaphysical fiction – *Gaia*, the Earth Goddess, Mother Earth – a romantic illusion and softening myth for the ontologically vulnerable. In questioning this claim about the modern and its contrary, I shall use as a guide the reasoning of none other than Max Weber himself.

As we have seen, Weber makes the point that the need for salvation arises in an urgent and compelling way from a conflict between a society's 'rationalised' image of the world and what people experience in the face of the 'actual' world. The conflict produces something 'specifically senseless' in the 'actual' world from which people wish to be redeemed. Weber discusses this issue in the context of religion – a metaphysical system that constructs a 'rationalised' image and at the same time promises salvation from the 'senseless' that it itself produces. Yet not all metaphysical systems are of this type. They may be 'rationalised' but not necessarily imbued with a salvation intent. At any rate, the point here is that if they are redemptive, this is something that cannot be taken for granted but must be demonstrated. Since Durkheim (1984 [1893], 1976 [1915]), a reliable index in this respect has been the extent to which the 'senseless' in the world is tolerated or, another way of saying the same thing, whether – as in the case of the nationalist's nation – an assault on the 'rationalised' image, whether symbolic or otherwise, is perceived as a sacrilege. These considerations set the agenda for the discussion on 'the disenchantment of the world' that follows.

The first step is to establish the metaphysical status of the modernist 'rationalised' image of the social; the second, to examine what, on the basis of this image, is experienced as 'specifically senseless' in everyday life; and the third, to explore how the 'senseless' is dealt with. If the modernist subjectivity is prepared to live without the ontological security provided by softening myths and illusions, if it has 'matured' enough to have no need for metaphysical crutches and 'guardians' and walks unaided by itself, as Kant had hoped, we would expect the 'senseless' to be more or less tolerated. This would provide credence to the modernist claim that those 'immature' among the 'mature' who cannot bear 'the fate of the times', like environmentalists, turn for protection to pre-modern metaphysical fictions or invent new ones. If this proves to be the case, we would have to acknowledge also that environmentalism constitutes a radical break with the modernist paradigm not only in epistemological but also ontological terms. In what follows, I shall try to show by contrast that the modernist logic of the Same has produced one of the most widespread and enduring metaphysical fictions about the nature of social reality; and that the modernist subjectivity has proved itself highly intolerant of the 'senseless' wherever and whenever it was encountered. This metaphysical fiction is none other than humanity and its corollary: humanism.

Humanism, it is often argued, is notoriously difficult to define. As Davies (1997: 2) suggests in a recent study, the problem has to do with its complex history and the 'unusually wide range of possible meanings and contexts'. Accordingly, the aim of his study is not to provide a single 'stable "meaning"' for the term, indeed, not 'even a range of sharply-focused definitions' (1997: 5). Rather, Davies resigns himself to exploring the complexity itself, and when he does try to draw some sort of general conclusion, the argument becomes tautological: 'what they share – what makes them all "humanisms" – is their conviction of the centrality of the "human" itself' (1997: 20). A much older definition fares a bit better. 'Humanism ... may be described in general as the attitude of mind which seeks the key to the world in the life of man or, at any rate, the key to man's life within himself' (Mackenzie 1907: 14). Mackenzie contrasts humanism with 'naturalism', which seeks the 'key' to 'man's' life in the world around him, and 'supernaturalism', which seeks the same key in powers beyond the world. Yet there are complications here too. The separation of the three notions, Mackenzie points out, is stable only insofar as humanism is defined in rather purist terms, in which case the notion undermines itself from within.

Humanism may be understood, in the first place, as the study of human life – and this is the least complicated but also the least comprehensive definition. It could also be understood to mean that 'the world as a whole is to be interpreted from the human standpoint' (1907: 186). The latter definition, in turn, could mean what Kant, Hegel and the other German idealists meant, namely, that the world is 'man's' creation, the product of the human mind – a phenomenon. But it could also mean that the world 'must ultimately be interpreted in relation to human life', even though it has an independent existence. Mackenzie himself prefers the latter interpretation but as he points out, in this watered-down version, humanism 'almost ceases to be opposed to naturalism' (1907: 187). It becomes a strange hybrid, one might add, a neither/nor attitude, a studied ambiguity that harbours an internal contradiction, the kind of subterfuge currently popular in the social sciences (for example, Bourdieu 1990). Human beings cannot be understood independently of the world in which they live – 'naturalism' or 'objectivism' in Bourdieu's terms – but neither can they ever be reduced to it ('humanism' or 'subjectivism').

The stronger version of humanism has its own problems. As Mackenzie points out, it could lead to the paradoxical situation where it is no longer 'truly humanistic' (1907: 188). The assumption on the basis of which 'man' becomes the measure of all things, the creator of the world and the only relevant Subject is one small step away from treating the world as a domain without intrinsic significance, with no value in and of itself. When this step is taken, when the world is perceived to have value only insofar as it serves human needs and interests, humanism undermines itself from within. It becomes less than human.

> A humanitarian is not one who constantly distinguishes man from all other beings and is proud of his superiority, but rather one who constantly recognises that the lower animals are his poor relations. ... Man ... is not a beast but ... [he is] most bestial when he forgets his kinship with the brutes, and with the general system of nature from which he springs. The attempt to live a life in which the animal nature of man and the external conditions of his existence are set aside, nearly always leads to a result which is strictly inhuman. (Mackenzie 1907: 188–89)

Although this text was written almost a hundred years ago, it could easily pass for an environmentalist critique of the modernist paradigm. If it had never become such a critique, it is only because of 'bad cultural timing', an unfavourable cultural context – which is to say also, if it needs to be said again, that 'facts' acquire relevance, significance and gravity under determinate cultural conditions.[7] This is precisely the kind of humanism that environmentalists themselves have in mind when they criticise modernist anthropocentrism – the primacy of the human – which they dismiss as the 'arrogance of humanism'. Yet the primacy of the human vis-à-vis the rest of the world does not exhaust humanism. The human is also primary vis-à-vis itself, when it encounters itself in other subjectivities. This kind of humanism, which is concerned with the humanisation of the human rather than the natural, is far less contested and for the purposes of the present discussion far more important.

Before I turn to examine the humanisation of the human, it may be pertinent to point out that the problems of definition discussed above stem largely from the contradictory status of the modernist subjectivity, which pretends to be both the creator of the world and a creature in it. This contradictory status generates a range of possible scenarios. As we have seen, in order to make a decision about what exists in its entirety and hence guarantee the truth of the real, 'man' has to step outside the world and assume a metaphysical position. He must become the Subject of the world, and it is this understanding that is reflected in the stronger version of humanism of the German idealist tradition. What is significant to note, however, is that this metaphysical position is also the condition of possibility of 'naturalism' itself. To say that 'man' is the product of the world in which he lives is to posit a double persona: 'man' as the product of the world and 'man' as the creator of the world, which in turn produces 'man'. To put it another way, the 'man' who makes this statement cannot be 'man' in the first sense. If he were the product of the world, his knowledge would be wholly determined by the world as the 'naturalist' rule postulates, in which case he would be completely unaware that such a rule existed in the first place. Being aware of this rule demonstrates that he is not merely the product of the world but also its creator (and in a roundabout, mediated way of himself as well). This is to say that if 'naturalists' – or 'objectivists' in the social sciences – can assert the dependence of 'man' on the world, it is only because they have effaced from their memory the inaugural ('subjectivist') act of creating a world on which 'man' would become dependent. It is to say also that 'naturalism' is a humanism in disguise or, if one prefers, in denial. Between these two extremes – humanism and naturalism, 'man' as the Subject of the world and 'man' as an object in it, 'objectivism' and 'subjectivism' – lies the epistemological position favoured by Mackenzie and much of current social theory – a position of half forgetfulness and half recollection that tries to mediate (or hide) an irreconcilable contradiction.

The second kind of humanism, which is my main concern here, is about the encounter between 'man' and himself and is primarily ethical in nature. Although both kinds of humanism have the same origin – historically the eighteenth century, ontologically the emergence of 'man' as Subject – and often complement one another, they are not always compatible. In general, humanists in the first sense – those who believe in the powers of 'man' and his dominant position in relation to the rest of the world – are also humanists in the second; they too posit the essential unity of humanity.[8] Yet the reverse is not always the case. Anti-humanists do not thereby cease being Humanists. The humanisation of the human is a much more fundamental, widespread and enduring modernist attitude than the humanisation of the natural. As Soper (1986: 128) points out in her discussion of structuralism – by all accounts, the contemporary anti-humanist discourse par excellence – it 'exemplifies the tendency of anti-humanist argument to secrete humanist rhetoric. For where, if not from humanists sources, does the language of struggle, victimization and loss originate?' Where else if not from humanist sources, indeed? Let us bear in mind that the man who put structuralism on the agenda, Lévi-Strauss, was an anthropologist and that, as I argue elsewhere (Argyrou 2002) and will return to below, in the vein of all anthropology, his fundamental concern was to demonstrate the essential unity of humanity, the underlying Sameness of the West and its Others.[9] Similar considerations apply to radical environmentalists whose anti-humanism is often said to be anti-Humanistic. Even Heidegger, the personification of twentieth-century anti-humanist philosophy and the main influence on deconstructionists of the likes of Foucault and Derrida who claim to have finally de-centred 'man', felt the need to put the record straight. In his 'Letter on Humanism', Heidegger complains that his anti-humanism is all too often associated with the anti-human. 'Because we are speaking against "humanism" people fear a defence of the inhuman and a glorification of barbaric brutality' (1977b: 225).[10] But such is not the case, Heidegger says in his defence. His philosophy, he argues, actually elevates 'man' to a much higher status than modernist humanism. The latter conceives of 'man' as a mere Subject, an intelligent animal. Heidegger assigns him the role of that unique opening through which Being manifests itself in the world.

Let us turn, then, to explore in greater detail this type of Humanism of which even the most anti-humanist of philosophers are, if not necessarily proponents, certainly mindful. As Davies points out, although this sort of Humanism is usually said to have emerged in the Italian city-states of the Renaissance, it is in fact a nineteenth-century construct. 'Renaissance man' was the invention of nineteenth-century scholars, such as the Swiss historian Jacob Burckhardt, and a reflection of how they themselves understood the meaning of humanity. For these nineteenth-century scholars, Davies argues, the essence of being human was:

> A universal capacity to think of yourself, in a fundamental way, as an individual: not as Florentine or Marseillais or a sailor or Roman Catholic or somebody's daughter and grand-daughter, important all those affiliations might be, but as a free-standing, self-determining person with an identity and a name that is not simply a marker of family, birthplace or occupation but is 'proper' – belonging to you alone. (Davies 1997: 16)

This type of subjectivity, then, is enmeshed in all sorts of particular affiliations. It may be a man or a woman, Italian or French, Catholic or Protestant, a sailor or a scholar, someone's daughter or granddaughter. What makes it the sort of individual that the Humanists of the nineteenth century had in mind is the ability to disengage itself from all particularity and circumstance and to think of itself as a unique subjectivity, as someone irreducible to any particular time or place. Yet although the individual learns to experience the self as a unique being, the experience itself is 'universal'. This is to say that, paradoxically, uniqueness becomes common – a common denominator – what above anything else unique individuals recognise in each other and hence what unites and makes them the same in their individuality and difference. 'Humanity is both general and special, common and rare. Each of us lives our human-ness as a uniquely individual experience; but that experience … is part of a larger, all-embracing humanity, a "human condition"' (Davies 1997: 21).[11] Let us call this paradoxical difference that makes the same 'Pure Humanity', and let us use 'Pure' to mean two things. First, what it meant for Kant in such expressions as 'pure reason', 'pure intuitions', and so on, namely, that which is uncontaminated by the empirical. As I have already suggested, my contention is that the epistemological conditions of possibility of Pure Humanity are structurally the same, and that the notion is as metaphysical as Kant's own 'pure' ideas. Second, and not unrelated to the first, let us use it to mean what it meant for Durkheim and his followers: the sacred as opposed to the profane, the collective as opposed to the individual and the individualistic, unity as the very embodiment of social innocence. This as an intimation of what is to be subsequently discussed further and demonstrated, namely, Pure Humanty's salvational intent.

The question that I am raising once again, then, is this: where does one need to go to see, not the Florentine or Marseillais, the sailor or the Roman Catholic, somebody's daughter or granddaughter, but over and above them that which transcends them and in transcending unifies them, namely, Pure Humanity? As I pointed out in the discussion on the 'World Picture', the fundamental effect of looking at the world from an external position is the equalisation of the beings of the world. This is what seems to happen in this case too. Concrete individuals, historically and socially situated, are equalised by means of an abstract, ineffable, universal quality – human-ness. Instead of repeating the analytical moves of the previous section, however, I propose to answer the question by making use of what has already been established: that stepping outside the world inevitably leads to contradiction of the sort that, as we have seen, Kant warned against but he himself could not avoid. What I propose to do, in other words, is to examine the contradictions embedded in the notion of Pure Humanity and use these as an index of its metaphysical nature.

To return to Davies, having discussed the numerous and not always compatible uses of Humanism in various nineteenth-century discourses, he raises the question as to whether the notion can be said to mean anything at all.

> Is it any more than a blank screen onto which anyone can project their flickering fantasies of power and happiness? It may be, of course, that it is precisely this protean adaptability and serviceable vagueness that gives the word its rhetorical power and range. For in these nineteenth-century discourses, the figure of the human, though deployed in contexts that might suggest that it is geographically and historically specific (European and modern), in reality signifies something that is *everywhere* and *always* the *same*. (Davies 1997: 24; my emphases)

Let us note the discrepancy here. On the one hand, the figure of the human is everywhere and always the same. This is reality – 'in reality'. On the other hand, some nineteenth-century discourses – one presumes colonial discourses – used it in certain 'contexts' which 'might suggest' that this figure was in fact modern European 'man'. Hence, the reference to the 'protean adaptability', 'serviceable vagueness' and 'rhetorical power' of the notion, since these qualities allow it to contradict itself without causing itself too much damage. Let us also note that this paradoxical state of affairs has something do with the nature of the 'reality' in which Pure Humanity becomes conceivable and the nature of the 'contexts' in which it is deployed, and that the ontological status of both is precisely what needs to be decided. Let us note, finally, that Davies does not appear to be totally convinced by the rhetoric of Humanism. 'We might call this the myth of essential and universal Man: essential, because humanity – human-ness – is the inseparable and central essence, the defining quality, of human beings; universal, because that essential humanity is shared by all human beings, of whatever time or place' (1997: 24). 'We might call it a myth': there are good empirical reasons for calling the notion of 'essential and universal man' a myth without so much reservation.

Davies' discussion moves on to the origins of Humanism to note that these were political rather than philosophical and lie in revolutionary discourses on rights. Among such discourses, he mentions Rousseau's *The Social Contract*, Paine's *The Rights of Man* and Jefferson's *Declaration of Independence*. 'Of course', Davies goes on to say, '"universality" is a tricky notion'. It is 'tricky' because 'universals may not always be quite as generously inclusive as they would have us suppose'. Here, then, is another discrepancy.

> It does not seem to have occurred to Jefferson and colleagues to extend the universal freedoms of the *Declaration* to their own or their neighbours' slaves. Mary Wolstonecraft's response to Paine … was to issue a *Vindication of the Rights of Woman* (1792) that his own book appeared to have overlooked. And Karl Marx pointed out that the heady rhetoric of 'Universal Man' that accompanied the revolutions of the eighteenth and nineteenth centuries tended to give way, once its ideological work was done, to a rather narrower and more pragmatic set of class interests. (Davies 1997: 26)

Universals are universalising, which is to say, they include all. Yet sometimes they compromise themselves and exclude certain parts of humanity. Paradoxically, universal rights were not so universal after all. At the time of their invention and for a very long time afterwards, they were not extended to blacks, women or the working classes. Davies, therefore, concludes on a note of suspicion. Although, he says, 'one of the effects of a universalising notion like "Man" is to dissolve precisely such particularities as race, sex and class … it is always prudent to ask what specific historical and local interests may be at work within [such] grandly ecumenical notions'.

Let us, then, explore in detail the paradoxes that Davies notes but does not resolve. But first, a quibble. The dissolution of particularities like race, sex and class is not 'one' of the effects of universalisation, as Davies says, but *the* effect, the most fundamental outcome of looking at the social world from a position beyond the world – which raises with more urgency the question as to why the dissolution of such particularities is followed by their partial reinstatement. What is one to make of the 'trickiness' of universality, which Davies mentions so casually? How can the notion of 'Universal Man' be both 'reality' and 'myth', both truth and ideological fiction? Why did it not 'occur' to Jefferson to extend universal freedoms to slaves? Why did Paine 'overlook' the rights of women, and why does the rhetoric of 'Universal Man' 'tend' to give way to more 'pragmatic' class interests? We cannot hope, of course, to know what Jefferson, Paine or the bourgeoisie of the eighteenth and nineteenth centuries were actually thinking when declaring universal freedoms and rights. Yet we do not need to know their inner thoughts to make sense of these paradoxes. The failure to uphold the universal has less to do with intention than with structural impossibility. It is in the ontological conditions that produced 'man' that we must locate the 'trickiness' of universality, not the thoughts and intentions of particular individuals. The universal is 'tricky', then, precisely because the modernist subjectivity strives to conjure up an impossible 'trick', namely, to constitute and reproduce itself as the Creator of the world. It is unstable because the modernist subjectivity is trapped in an ontological double bind, the interplay between being both Subject and object, both the guarantor of the truth of Pure Humanity and the human – 'only too human', says Nietzsche – who must live this truth in a less than pure world.

It is worth distinguishing two levels of analysis here. The first is phenomenological. Although the modernist subjectivity is able to step outside the world and imagine it as Pure Humanity, that is, although it can symbolically deny determining value to particularity, circumstance and difference, it cannot maintain human purity forever. Sooner or later, it must return inside the world and take its place in it as a socially and historically situated person, a concrete individual among other Florentines, people from Marseille, sailors, Roman Catholics, somebody's daughters and granddaughters. This means at least two things, both of which Davies notes. It means first, that the modernist subjectivity is forced by the pressures and urgencies of the social to give in, whether unwittingly or unwillingly and reluctantly, to 'pragmatic' individual or collective interests. It means also that because it has never actually stepped outside the social world and hence has never been a totally free-floating subjectivity, its thinking remains subject to the prevalent cultural conditions, which produce blind spots and render certain things unthought and certain things unthinkable. These are the things that, given the cultural conditions of the time, were 'overlooked' by Paine – the rights of women – and 'did not occur' to Jefferson – liberation of slaves. Hence, the compromise of Pure Humanity in the figure of 'man' as opposed to woman, and the white, bourgeois, modern European as opposed to black, slave, proletarian and traditional or primitive Other respectively. Pure Humanity is the partly context-free reality (the 'in reality' that Davies mentions) of the modernist subjectivity as Subject; white, bourgeois, modern, European, man, are some of the social and historical contexts in which the modernist subjectivity operates as object – the contexts in which it compromises both itself as Subject and the social innocence of human purity.

This interplay between being Subject and object, incidentally, goes a long way in explaining the 'ambivalence' of colonial discourse that postcolonial scholars often note only to reduce to psychologistic explanations (e.g. Bhabha 1994). Moreover, as it is now increasingly recognised in the anthropological literature but not adequately explained, it makes colonialism itself neither wholly a civilising mission – an idealistic attempt to put Pure Humanity into practice – nor wholly an imperialist ploy – a cynical recognition that Pure Humanity is impossible in an impure world. This interplay, finally, makes Pure Humanity neither 'reality' nor 'myth' but both things at the same time, a 'mythical reality', as Sahlins (1985) might say – a reality experienced in mythical time and a myth lived in everyday life as reality.

Yet the impossibility of Pure Humanity cannot be wholly explained by the inevitable compromises that the modernist subjectivity must make in the course of everyday life, whether unwittingly or unwillingly. At a more fundamental level than the phenomenological, it is compromised long before the need arises in practice. It is *always already* compromised, at the very instant it is visualised and thought. This is to say, among other things, that what is compromised in everyday life is not Pure Humanity – one is tempted to say 'not Pure Humanity as such', which would be meaningless – but its semblance, a simulacrum or, if one prefers, the phenomenon that goes by that name. We have already encountered the paradox of positing the impossible, of which Pure Humanity is but one manifestation, in Kant and the thing-in-itself, which cannot be known but must exist if anything else is to exist. It runs through all of Western philosophy and is encountered even in attempts to expose it, as in Derrida who, in an attempt to demonstrate the impossibility of 'presence' (the thing-in-itself or Being), ends up making its absence present. 'The final paradox of the search for purity', Douglas (1966: 162) points out in another but not unrelated context, 'is that it is an attempt to force experience into logical categories of non-contradiction. But experience is not amenable and those who make the attempt find themselves led into contradiction'. My aim here is not to engage in further discussion of these well-known ideas but simply to provide an anthropological example of the impossibility of Pure Humanity which I explore in greater detail elsewhere (Argyrou 2002). I should also say that my intention is not to make an ontological statement about this impossibility. The inability to know the thing-in-itself, and the interest in the issue that makes it an issue, cannot be reduced to anything like the nature of the human condition. It reflects, rather, the assumptions of a particular culture, at a particular historical conjuncture, in a particular part of the world.

It is no accident that twentieth-century anthropological Humanism expresses itself primarily in the negative instance. It does so because it is impossible. And it is impossible both because it must be compromised at the phenomenological level and because it is always already compromised before it must. Anthropological Humanism revolves around the negative notion of ethnocentrism, the idea that Western cultures are superior to the cultures of other parts of the world. For anthropologists this is definitely *not* the case, but the attempt to say what the case *is* exactly is fraught with intractable paradoxes. 'Cultural relativism', the only such attempt, was developed at the turn of the century precisely to counter the evolutionist claim that Europe occupies the pinnacle of cultural achievement.[12] Understood to mean that cultures are incommeasurable is highly problematical, not least because it leads to an untenable epistemological relativism. Moreover, as Evans-Pritchard pointed out long ago, if other cultures were completely different from those of the West, they would be unknowable. But it could also lead to an ethical relativism, which is even more problematical. Anthropologists would have to tolerate what they perceive as inhumane practices in other cultures – the mutilation of female sexual organs, for instance. If they did tolerate such violations, they would be compromising their Humanism. Women too embody that core essence – humanity or human-ness – that, as Davies says, is 'shared by all human beings, of whatever time or place'. Not tolerating such practices, on the other hand, does not fare any better. It leads, by a different route, to the same outcome: compromise of Pure Humanity. Cultures, much like individuals, are everywhere the same, not in form to be sure, but in content – in terms of cultural value and worth. They are the same in this sense because the individuals who produce them are everywhere and at all times the same. Hence, the frequently heard but not unjustifiable criticism that any attempt to impose one's ethical values on other people is ethnocentric. Such are the dilemmas that anthropologists face or, to be more precise, one of the ways in which they experience, enact and reproduce the impossibility of Pure Humanity. But, as I have already suggested, there is another level at which this impossibility manifests itself and this manifestation, which is not perceived by the anthropologist and in this sense constitutes the 'anthropological unconscious', is even more fundamental.

Consider the defining anthropological discursive practice – interpretation. The ostensible aim of any kind of interpretation, of course, is explanation. To explain is to make what is different and alien understandable, that is, in effect, as Nietzsche pointed out long ago, to make it the same. Anthropological interpretation takes the exotic and the alien, a particular practice, for instance, and by means of analogy renders it in terms that can be understood in the anthropologist's own culture. Yet this is not mere translation. Both the intent and the anticipated outcome of anthropological interpretation are ethical. The anthropologist's aim is to show that no matter how different and exotic other peoples appear to be, 'in reality' they are essentially and fundamentally the same as 'us'. This serves as a corrective to the ethnocentric assumption that because they are different, they are also inferior. The aim of anthropological interpretation, in short, is the redemption of Otherness. Take the paradigmatic case of Evans-Pritchard's interpretation of Zande witchcraft, a belief system that, as he argues, is not meant as an explanation of empirical reality. If someone is injured while walking in the bush, say by a buffalo, the Zande do not dispute this empirical fact. What they are trying to explain is the timing of the event – why this particular individual and not someone else, why then and not at some other time. We say, Evans-Pritchard says – and this is the analogy that makes the same – it was an 'accident'; the Zande say it was 'witchcraft'. Although the idioms used are different, both 'us' and 'them' refer to the same thing. The Zande, therefore, are rational people like 'us'. Yet as soon as this Sameness between 'us' and the Zande is asserted, at the very same instant it is invalidated by the structure of things themselves. The Zande are unaware that witchcraft is their way of making misfortune meaningful. They believe that it is a magical substance in the body of the witch. This 'failure' to grasp the 'true' meaning of witchcraft makes the Azande different from the anthropologist – different and, given that in the anthropologist's culture knowledge is valued more than ignorance, inferior. The anthropologist knows the 'truth' about witchcraft; the Zande, whose practice it is, do not. This is to say that to make them the same – rational people like 'us' – the anthropologist must posit a Zande unconscious – which makes them very much *unlike* 'us'. It is to say that the anthropologist becomes what he does not wish anyone to be – ethnocentric. Inadvertently but inevitably, he refutes what he sets out to demonstrate, namely, the essential and fundamental Sameness between 'us' and 'them' – Pure Humanity.

My aim in the foregoing discussion has been to establish the metaphysical status of Pure Humanity by following Kant's own reasoning and apodictic method – which, as we have seen, he posited only to betray – namely, that any attempt to venture into the metaphysical inevitably leads to contradiction. Kant sought to define the nature of physical reality but he could do so only metaphysically. To draw the boundaries of the real, he found it necessary to step beyond the empirical world into the realm of the imagination; to guarantee the truth of the knowable in nature, he was forced to posit the truth of the unknowable – of the thing-in-itself. This is to say, as Collingwood says, that he ended up 'flatly' contradicting himself. Nineteenth-century Humanists sought to define the nature of human reality and fared no better. They too could do so in no other way than metaphysically – in the cases explored, by venturing outside society and history and taking up the cosmic perspective of the Subject of the world. Consequently, they produced a reality that is not of this world but nonetheless must necessarily be lived in society and history in their capacity as objects: Florentines, Marseillais, sailors, Roman Catholics, concrete individuals who occupy certain social positions, are enmeshed in particular social relationships, think and act in culturally and historically specific ways. They constructed an image of a seamless human whole about which nothing social and historical can be known precisely because nothing can be distinguished in such terms. In short, they produced the social equivalent of Hegel's Pure Being, which, being so pure, so uncontaminated by the empirical, is nothing. Hence, Humanism's 'protean adaptability' and 'serviceable vagueness' of which Davies complains. Because Pure Humanity and its corollary, human purity, are socially nothing – for they have never existed and can never exist in society and history – Humanism can be moulded into almost anything.

This metaphysical vision is the modernist 'rationalised' image of the social par excellence and, on the basis of Weber's schema, the key to what the modernist subjectivity experiences as 'specifically senseless' in the 'actual' world. As such, it opens up an important domain of inquiry which Weber foreclosed by insisting that the modernist 'senseless' is the 'fate of the times' which this type of subjectivity, unlike subjectivities in other times and places, faces like the 'man' that he is. A discussion of this 'senseless' is fraught with difficulties however. To call Pure Humanity the key to it, for instance, is somewhat misleading. It pretends to make a discovery of something that anyone with a modernist sensibility has come up against repeatedly and in a certain sense knows quite well. The key, if there is one, does nothing more than unlock the context in which the experience may be interpreted in wider ontological terms. Then there is the danger of trivialisation. As I have pointed out several times already, my intention is to question the claim that apart from a few pockets of difference – the 'immature', the 'pathological', the 'traditional' – the West is a disenchanted domain. Yet 'disenchantment' is an old-fashion term of an old-fashion lexicon of a paradigm that, allegedly, has been superseded (by the 'postmodern') or is reinventing and revitalising itself ('reflexive modernisation'). How relevant can this notion be in the new order of things? This question cannot be answered in the abstract. The index of the notion's relevance is reflected by the extent of its use. As I have already suggested, although the term itself is not frequently encountered, the critique of such highly relevant discourses and practices as environmentalism and nationalism is largely made possible by the assumptions that underscore the notion of disenchantment. There is, finally, and perhaps more importantly, the critical issue of empirical evidence. In his aphoristic mode, Nietzsche noted that since it had killed God, 'man' has become fanatically moral. The argument developed here is that Pure Humanity constitutes the modernist ethical myth par excellence. How can this claim be verified?

One could, no doubt, do an ethnographic study of how the 'senseless' of Pure Humanity is experienced and dealt with in everyday life. Or a study of Humanism that would, among other things, examine the social conditions that account for the variable emphases on different aspects of this 'senseless' (or the invention of new ones). Or a comparative study of modernist Humanism and, say, the Humanism of Christianity. All these would be legitimate projects but quite beyond the scope of the present discussion, which is simply to outline the structure of a cultural logic. In what follows therefore, I restrict myself to noting the proliferation of discourse on the modernist 'senseless' – the persistence of older themes, the emergence of new ones, the shifts of emphasis and the shifts in perspective. I shall take this proliferation as a partial but nonetheless reliable index of the relevance, significance and gravity of this 'senseless', of the threat that it poses to the image of Pure Humanity, the concern about this threat of those concerned, in short, the degree of intolerance that the modernist subjectivity demonstrates.

The preceding discussion on Humanism should have provided more than a hint as to what the modernist 'senseless' might be. As Davies points out, the effect of the universalising notion of the Human is the dissolution of such 'particularities as race, sex and class'. Let us note what should be obvious: that these social divisions, and several others, constitute the 'face' of the 'actual' world that comes into conflict with the 'rationalised' image of Pure Humanity – the 'experience' that, as Douglas argues, cannot be forced into 'logical categories of non-contradiction'. How these divisions emerged, both in themselves and for themselves, is a sociological question that, as I have just pointed out, is beyond the scope of the present discussion. We can note, however, that all three have their origin in the nineteenth century, with social class being the most prominent theme. Let us also note that what is 'senseless' about these divisions is not their existence *per se* but precisely the fact that even though the modernist subjectivity dissolves them in the realm of the imagination – which it elevates to the status of reality as such – it is unable to do so in the course of everyday life. The modernist subjectivity does not deny that these divisions exist in practice. On the contrary, it constantly draws attention to them. What it *does* deny is that they are intrinsic to social reality, an inevitable part of the human condition. It is this prospect that makes them intolerable. As long as the modernist subjectivity can explain them away – as the result of, say, ignorance, arrogance or greed – and as long as it can act on the basis of these assumptions, the threat that such divisions pose to Pure Humanity can be contained and to a certain extent neutralised. But the inability to account for them, and their persistence despite every effort to educate the ignorant and to reform the arrogant and the greedy become menacing. They do because, as Geertz (1973: 108) points out in what, following the 'postmodern' ethnographers, one could call an 'ethnographic allegory', they 'raise the uncomfortable suspicion that perhaps the world, and hence man's life in the world, has no genuine order at all … that life is absurd and the attempt to make moral, intellectual, or emotional sense out of experience is bootless'.

Let us note, finally, taking a cue from another 'ethnographic allegory' – Douglas' (1966) work on purity – the relative status of these divisions in the wider economy of meaning and cultural value. Since the impure is the outcome of a society's particular system of classifications, modernist social divisions would cease being intolerable, even if still 'senseless', when, as result of shifts in the system, the value attached to them diminishes. Such seems to be currently the case with social class but apparently not with race or gender. As an extension of Douglas' argument, it may be worthwhile to point out also that shifts in the system of classifications could raise to the level of intolerance 'new' social divisions – divisions, that is, already in existence but not yet imbued with the necessary significance. Examples of such emergent divisions in the current climate of modernist sensibility are those based on sexual orientation, disability and age. This admittedly schematic discussion of the modernist 'senseless' would not be complete without mention of two other major divisions, as intolerable as those based on race and gender. The first is what anthropologists call the 'Great Divide' – the division between the West and its Others. In its guise as a struggle against racism, this division has its origins in Victorian anthropology, particularly the work of E.B. Tylor. As a struggle against ethnocentrism, it emerged at the turn of the century in the guise of an 'epistemic' critique of evolutionism. The second major division, which seems to have become intolerable during the second half of the twentieth century, is based on ethnic identity and manifests itself primarily in the form of nationalism. This is not to say any sort of nationalism. Nationalism becomes intolerable when it itself does not tolerate other ethnic identities. Much the same can be said about the anthropologist's 'Great Divide'. Cultural difference, which is ordinarily celebrated in the discipline as 'cultural diversity', becomes 'ethnocentrism' when it is denied the same cultural value and worth. In general, the guiding logic in dealing with social divisions is that empirical difference is tolerated only insofar as it can be explained away, when it can be shown that it has no determining value in the wider scheme of things.

To be sure, human unity has been part and parcel of religious imaginings both in the West, as in the Christian paradigm, and in many other cultures of the world. No other culture, however, has produced anything like Pure Humanity. In no other culture has the human become the focus of so much attention and concern, valorised more, elevated higher, accorded the status of the universal. Nowhere else has this happened before because in no other culture has the human – human-ness – become the object of its own worship. This is to say, among other things, that nowhere else has the social become the exclusive domain for 'ontological security' and meaning. Transcendence has been for the most part experienced as venturing outwards, as having an encounter with something other than the human. By contrast, having trapped itself in an 'internally referential system', the modernist subjectivity can transcend only insofar as it doubles itself. As a result, it encounters nothing other than itself and mistakes itself for the transcendental. It is impossible to resist noting another 'ethnographic allegory' in this respect. If there is any place in the world where the social is mistaken for God, this is not the 'elementary' societies that Durkheim had in mind but, on the contrary, the 'complex' societies of the modernist West.

If the foregoing discussion is anything to go by, one or two things begin to emerge in a rather different light. First, the ontological nature of the modern condition. As we have seen, the modernist subjectivity presents itself as a heroic, almost tragic persona – a persona that must bear the meaninglessness of the modern condition for the sake of human dignity, which for the modernist translates into individual autonomy and freedom. The best one can do under modern conditions is to find meaning in the private sphere – in romantic love, for instance, as Weber (1946) noted, or 'the pure relationship', as Giddens (1991) recently reiterated. 'Individuals seek the meaning of their existence outside of religion and politics', argues Ferry in the same vein. They have to. 'The question of the meaning of meaning can no longer be posed collectively within the heart of a secular universe'. Hence, Ferry reasons, the rise of various 'fundamentalisms' in the 'heart' of the modern, prominent among them environmentalism. Unlike romantic love and other private distractions, it is public and collective, a political ideology imbued with religious imaginings that fills the ontological void. But if so, what is one to make of Pure Humanity? It is as public, collective and political an ideology as any. In its various manifestations, it has been at the 'heart' of this 'secular universe' for almost two centuries. It has the same metaphysical credentials as any transcendental system. And it has proved itself as intolerant of 'impurity' and 'profanity' – class discrimination and exploitation, gender inequality, racism, ethnocentrism, nationalism and various lesser -isms – as any religion. Pure Humanity, it seems, has been the best-kept modernist secret – kept, that is, by the modernist subjectivity from its own, otherwise analytical gaze. It has to be. Cultural blindness is the condition of possibility of any myth worthy of the name.

The second thing that emerges in a different light is what often puzzles apologists of the modernist paradigm, like Ferry, about environmentalists. As we have seen, Ferry cannot make head or tails of a seeming contradiction in radical environmentalism. On the one hand, he says, there is 'the love for the native soil' and 'hatred of cosmopolitanism … and the universalism of the rights of man'; on the other, there is 'the dream of self management, the fight against capitalism, racism and neo-colonialism'. Denial of universalism seems to be combined with a struggle against various forms of human division. 'The same obsession with putting an end to humanism is being asserted in at times schizophrenic fashion, to the point that one can say that some of deep ecology's roots lie in Nazism, while its branches extend far into the distant reaches of the cultural left'. Yet there is no contradiction in environmentalism, much less a suppressed schizophrenic condition perceptible only in the more radical discourses, such as Deep Ecology. If it appears so, it is only because the apologists of the modernist paradigm do not fully understand the logic of the culture they champion. Environmentalists do not deny Humanism. They are far too modernist to tolerate any sort of human divisions. They are extending, broadening, widening, universalising more, totalising at an even grander scale. They are striving to efface the last 'Great Divide' of the modernist paradigm – that between humanity and nature – to unify the world at the most inclusive level possible, stretching in the process Humanism beyond itself so that it no longer recognises itself. Environmentalists do nothing more than to take the modernist logic of the Same to its logical and onto-logical extreme. With environmentalism, modernist culture becomes fully articulated and further entrenched.

Notes

1. R. Lee and I. De Vore, *Man the Hunter* (Chicago: Aldeno, 1972). In North (1995: 201).
2. For a debate on these charges among environmentalists themselves, see Zimmerman (1994) and Watson (1981).
3. The roots of the 'environmental crisis' have been traced back to Christianity (White 1967) as well as the scientific revolution of the seventeenth century (Passmore 1980, Merchant 1980).
4. For reasons that I explore elsewhere (Argyrou 2003), reflexive modernisation can be taken seriously neither as a social theory nor as a myth.
5. On this issue see Ariès' (1983) excellent historical study of perceptions of death in Europe.
6. Escobar (1995) raises the same point but he does not elaborate.
7. For an even earlier and better-known critique that could not have become 'environmentalist' for the same reasons (and to which environmentalist themselves often refer) see Bentham (1907) – admittedly an unlikely source, given his utilitarianism.
8. To avoid confusion, I shall from now on refer to the humanism concerned with the humanisation of the human as Humanism.
9. In *The Savage Mind*, for instance, where mythical thought is presented as 'the science of the concrete' – a science that has the *same* value and status as the Western science of the abstract.
10. It should be noted here that Heidegger was briefly a member of Hitler's National Socialist party. Perhaps this statement is also an allusion to accusations of being a-one-time Nazi.
11. Geertz (1973: 53) in a similar vein: 'To be human ... is ... not to be Everyman: it is to be a particular kind of man, and of course men differ'.
12. It is often said, of course, (e.g. Marcus and Fischer 1986) that cultural relativism was invented as an epistemological tool. Yet this is an empiricist claim. As anthropologists are well aware, epistemology, much like everything else, is mediated by cultural concerns.

4 'Beyond Humanism': and further to the other side

'The Religion of Humanity', the Religion of *Gaia* and other Homologies

The more radical among environmentalists strive to take the world 'beyond humanism'. They could be articulating a wider sentiment about 'man's' three-century-old reign. As I will argue below on the basis of a related historical precedent, they may be giving unmediated, systematic and unambiguous expression to a desire that could be more pervasive than usually assumed.

'Green politics', writes Charlene Spretnak in a classic statement of the radical environmentalist position, rejects humanism. It rejects the 'philosophy which posits that humans have the ability to confront and solve the many problems we face by applying human reason and by rearranging the natural world'. As we have seen, for Spretnak and like-minded environmentalists, 'it is *hubris* to declare that humans are central figures of life on Earth and that we are in control'. It is hubris and an illusion, 'a dangerous self-deception contained in both religious and secular humanism'. 'In the long run, *Nature is in control*' (Spretnak 1984: 234).[1] Not 'man', then, but nature is in charge. Let us note in passing – not because it is a trivial matter but because I shall return to it below – the grounds of 'self-deception' and the conditions of possibility of 'truth'. They appear to be temporal: '*In the long run*, Nature is in control'. Modernist 'philosophy', according to this argument, misconstrued the relationship between human beings and nature because it contemplated the world from a rather narrow temporal perspective – the short run. In the wider scheme of things, however, it is nature that controls our destiny. Modernist 'philosophy' is by all accounts three centuries old. How long, then, must the 'long run' be to avoid the trap of arrogance and self-deception? How much time does the truth about reality require to emerge into the open? More importantly, where does one need to go to see that nature is control when, as environmentalists themselves insist, we are very much part of nature, not over and above it? What sort of disentanglement is this, and how is it possible?

But to raise these questions is to run ahead of the story. Let us for the time being return to 'The Spiritual Dimensions of Green Politics' and its call for transcendence in its full version – 'Beyond humanism, modernity, and patriarchy' (Spretnak 1984: 234). 'Green politics goes beyond not only the anthropocentric assumptions of humanism but also the broader constellation of values that constitute modernity'. What are these values? They are, the 'mechanistic analysis and control of human systems as well as Nature, rootless cosmopolitanism, nationalistic chauvinism, sterile secularism, and mono-culture shaped by the mass media'. Spretnak contrasts the 'values of modernity' with what environmentalists strive to achieve. 'An enthusiast of modernity has little use of the traditional institutions that further human bonding – the family, the church, community groups, ethnic associations' (1984: 235). Enthusiasts of modernity may not care much about 'human bonding' but enthusiasts of the environment apparently do. As Zimmerman (1994: 6) points out, 'many radical ecologists envision the emergence of nonauthoritarian, nonoppressive, non-hierarchical, "postmodern" societies in which free, playful, decentered, heterogeneous people live in small, bioregionally oriented … communities'. People living together in harmony in small, face-to-face communities, leading free and playful lives on the basis of something other than 'mechanical solidarity' – Zimmerman, at any rate, imagines them to be both 'decentred' and 'heterogeneous'. Finally and as importantly, 'Green politics' strives to take the world beyond patriarchy. 'The third cultural force that Green Politics counters is patriarchal values. In a narrow sense, these entail male domination and exploitation of women'. In a broader sense, they entail 'love of hierarchical structure and competition, love of dominance-or-submission modes of relating, alienation from nature, suppression of empathy and other emotions.' (Spretnak 1984: 236). 'Green politics', then, envisions a 'post-humanist, post-modern, and post-patriarchal society' (1984: 238) – a new society, to be sure, nothing less than a change in 'civilisation'. What is one to make of this vision of the world?

We must return to Soper and her astute observation that discourses such as these 'exemplify the tendency of anti-humanist argument to secrete humanist rhetoric', that the language of struggle, victimization and loss they use originates in humanist sources. In the interest of precision and clarity, we must return also to the distinction drawn in the last chapter between different types of humanism and restate Soper's argument to reflect it: 'anti-humanist arguments tend to secrete Humanist rhetoric' – the critique of 'man' as the measure of all things often goes hand-in-hand with the critique of human divisions. Victimisation, loss and redemptive struggle are key themes here. Beyond the victimisation of nature and the consequent environmental destruction, the modernist paradigm is responsible also for the victimisation of humanity. To begin with, it treats human systems 'mechanistically' and tries to exert 'control' over them. Yet human beings are not machines and should not be the subject of mastery. The modernist paradigm is responsible also for humanity's general state of alienation. Human beings have been alienated from nature and from one another through modernity's patriarchal structures, 'rootless cosmopolitanism', 'nationalistic chauvinism' and its general disregard for 'human bonding'; from wider, transcendental realities through its 'sterile secularism' and from their unique selves by means of the 'mono-culture' produced by the mass media. The modernist paradigm is responsible for all these human losses and more. Hence, the language of victimisation, redemptive struggle and radical transformation, and the vision of a post-humanist, postmodern utopia.

It may be argued, of course, that the humanism which environmentalist discourse 'secretes' is of a different kind from the Humanism of the modernist paradigm or, at any rate, that although both are essentially about human 'bonding', this vision of unity is achieved through different means. Modernity's Humanism, it could be argued, is rationalistic and abstract – in Weber's terms, an intellectualised, 'rationalised' vision of the world; that of environmentalism is concrete and experiential, generated in particular contexts and theorised on the basis of accumulated experience. As we have seen, and will return to again below, such is precisely the argument that ecofeminism uses in distinguishing between its own kind of identification with non-human beings and that of other radical ecologies. But there is a serious problem with this sort of empiricist argument, which I have raised several times already. This is the problem of the 'facts' that speak for themselves and mean by themselves – unmediated, independent of wider cultural assumptions about the nature of reality and what it means to be. I shall not reiterate the argument here. I shall turn instead to an ecofeminist attempt to ground facts in themselves and the problems that this and any other effort of this sort encounter.

'Feminist *ethics*', writes Karen Warren, 'rejects attempts to conceive of ethical theory in terms of necessary and sufficient conditions, because it assumes that there is no essence (in the sense of some transhistorical, universal, absolute abstraction) of feminist *ethics*'. Mainstream ethical theory fell into the trap of positing such an 'abstraction'; feminist ethicists are careful not to. Nonetheless, they too find it necessary to note some necessary conditions, which Warren prefers to call '"boundary conditions" of a feminist *ethic*'. Do these conditions imply an 'absolute'? They do not, according to Warren. They are simply meant as a means of clarifying themselves! 'These boundary conditions clarify some of the minimal conditions of a feminist *ethic* without suggesting that feminist *ethics* has some ahistorical essence'. Warren goes on to clarify further. 'They are like the boundaries of a quilt or collage. They delimit the territory of the piece without dictating what the interior, the design, the actual pattern of the piece looks like. Because the actual design of the quilt emerges from the multiplicity of voices of women in a cross-cultural context, the design will change over time. It is not something static' (Warren 1998: 335; my emphases).

In a typical empiricist fashion, Warren struggles to produce a historical, contextual, experiential ethical system. This is the 'interior', the 'design', the 'actual pattern' of the 'quilt'. But she cannot do so without a structure for the system, the 'minimal conditions', the 'boundaries' in which the pattern will be set, what defines the quilt as a quilt and the design as a quilt design. She cannot do so, in other words, without the wider cultural assumptions that render instances of female oppression visible, relevant and meaningful – an ethical issue to be taken up and brought to public attention. She must set these boundaries and deny them at the same time – set them, that is, *a priori* and then deny that they are ahistorical and universal. Hence, the tautology about the 'conditions' – the 'boundary conditions' that clarify the 'minimal conditions' – and the uneasy to and fro between a feminist *ethic* (the boundaries that 'delimit') and feminist *ethics* ('the design that changes over time'). This uneasy oscillation is nothing other than a manifestation of the need of the modernist subjectivity to double itself – if is to reproduce itself – the inevitable interplay between being the Subject of the world and object in it. Warren goes on to outline the boundary conditions of feminist 'ethics' but the essence of the argument has already been stated a few pages earlier: 'While feminists disagree about the nature of and solution to the subordination of women, all feminists agree that sexist oppression exists, is wrong, and must be abolished' (Warren 1998: 326). The pattern of the 'quilt' may change over time – notice, incidentally, the 'anthropologically correct' reference to a 'cross-cultural' context, which imposes, at least partly, the need to deny an ahistorical essence in feminists ethics – the pattern, then, may change but the understanding that sexist oppression is *wrong* and must be abolished cannot. If it did change, feminism would no longer be itself. Nor can this understanding be anything other than ahistorical in its conception and universal in its application. It is ahistorical because there is no one who can ground it in society and history; and it is universal because no instance of sexist oppression can be discounted, irrespective of whether it is 'inter-' or 'cross-cultural'.

The division between men and women is one manifestation of the 'senseless' that undermines the modernist 'rationalised' image of the social world – the vision, that is, of Pure Humanity. As we have seen, this division is 'senseless' not because men and women are the same in the common sense of the term but because men use empirical difference to deny women the same value and worth – their full Humanity or Human-ness. This is 'wrong' and must be 'abolished'. It is and it must because 'in reality' – which is the reality of the 'unseen', as William James (1961 [1907]) would say – men and women are essentially and fundamentally the same. There are more 'minimal conditions' than the denunciation of sexism, more manifestations of the 'senseless' that the ecofeminist and more broadly environmentalist ethic shares with the ethics of the 'enthusiasts of modernity'.

Ecofeminism, says Warren, differs from traditional feminism because it shows that the subordination of women is based on the same logic as the subordination of nature. It 'reconceives feminism' because it shows that *all* forms of oppression are based on the same patriarchal logic – the logic of domination. 'Other systems of oppression (e.g., racism, classism, ageism, heterosexism) are also conceptually maintained by a logic of domination. ... By clarifying this conceptual connection between systems of oppression ... [ecofeminism] leads to a reconceiving of feminism as *a movement to end all forms of oppression*' (Warren 1998: 330). A grandiose claim, one might say. Perhaps. Yet it is not any grander than the much earlier Marxist claim – that the end of class society will bring an end to all forms of domination – and does nothing more than reiterate the same desire. This homology should not be surprising. Ecofeminism, and more broadly environmentalism, is inspired by the same mythical vision of the social world – Pure Humanity, which, as we have seen, is inextricably a vision of human purity and social innocence.

It is because environmentalists share this metaphysical vision of the social, because they are as Humanists as the 'enthusiasts of modernity' despite their anti-humanism, that they find the 'senseless' which manifests itself in human divisions 'wrong' – indeed, as wrong as the 'senseless' that divides humanity and nature, which is their focal point. 'Thinking of nature alone will not do', Gottlieb (1996b: 525) reminds those environmentalists who neglect the 'senseless' in society. 'We are creatures of history and society as well as of the earth and air. We hunger for justice as we hunger for food and water. And without a compelling memory of class, gender, racial, and national forms of oppression ... deep ecology ... will be blind to complex and painful issues of social injustice and political reorganization'. 'Hunger' for social justice and 'pain' for the 'senseless' in society: environmentalists find social divisions as intolerable as any modernist subjectivity. They do so because they too are modernist subjectivities and share the same ontological predicament. The 'senseless' that undermines Pure Humanity must therefore be 'abolished' in *all* its forms and manifestations. It constitutes a scandal of the highest order, a profanity that threatens the purity, innocence and meaning of the social world.

Yet it is also true that environmentalists are not Humanists pure and simple. As I have pointed out several times already, the environmentalist vision of the world is more objectified, totalising and unifying than its modernist counterpart. It is mindful of Pure Humanity but only as an instance, albeit an important one, of a wider and far more unifying reality – so unifying, in fact, as to leave nothing outside. The environmentalist perception produces a vision of Pure Being, a resurrected version of Hegel's thing-in-itself or, at any rate, a vision in which, as a bare minimum, the common sense division between humanity and nature is not thought to be either critical or determining of the 'true' nature of things.[2] Environmentalists are more than Humanists, so much more that they are often misconstrued, and sometimes misconstrue themselves as anti-Humanists.[3] Over and above 'human bonding', there is the equally necessary bonding between the human and the non-human to be achieved. They carry a heavier ontological burden than the 'enthusiasts of modernity'. Over and above the 'senseless' in society, there is the 'senseless' that society has produced by alienating itself from nature to be exorcised. It should come as no surprise, therefore, that there are numerous homologies between the modernist and environmentalist ethical systems, that terms, ideas, arguments, justifications bear close resemblance and that to the extent they differ, this is largely in terms of semantic elasticity and inclusiveness. Nor should it come as a surprise that practices reproduce themselves faithfully over the course of centuries, despite differences in objectives. Unaware of it, activists, past and present, follow the same inexorable cultural logic. Finally, it should come as no surprise that environmentalist ethics faces the same intractable problems that plague the ethics of Humanism.

There are several key notions that environmentalists borrow from the modernist ethical arsenal to articulate their vision of the world. All are semantically expanded. They have to be. The task at hand is none other than the critique of the grandest of all modernist divides – the division between humanity and nature. Hence in the place of all the –isms that the modernist subjectivity uses to criticise divisions among human beings, we find –isms that contain and reproduce them at a higher register. 'Anthropocentrism', for instance – the view that human beings are the measure of all things – is structurally homologous to ethnocentrism, sexism, heterosexism, nationalism, classism, ageism and whatever other –isms are currently in use. It denies human superiority over non-human beings on the basis of the same logic that denies the cultural superiority of the West, of men over women, of one sexual orientation, nation, class or age over others. There are more homologies of this sort. Racism, as opposed to ethnocentrism, explains cultural superiority as the result of natural differences in ability. But what is one to call a similar attitude on a much larger scale – discrimination against other species of beings? Environmentalists opted for the most logical, if not most eloquent term: 'speciesism'. In Humanist ethics, the usual response to racism is denial of differences in natural abilities. In his critique of 'speciesism', Singer considers this argument and proposes a different line of attack, one that covers both cases of discrimination as well as the related case of sexism. 'Equality', he says, 'is a moral ideal, not a simple assertion of fact' (Singer 1998: 29). The implication of this principle is that 'our concern for others ought not to depend on ... what abilities they have. ... It is on this basis that the case against racism and ... sexism must ... ultimately rest; and it is in accordance with this principle that speciesism is also to be condemned' (1998: 30). Not natural ability, then, but something else should be the criterion of equality. Singer resurrects a principle from Humanist ethics, the nineteenth-century utilitarian principle of suffering. 'The question', Bentham said in his defence of

'the rest of the animal creation' to an audience that could not have been receptive
– a case of bad cultural timing, as we have seen – 'the question', then, is not
whether a dog or a horse or any other animal can 'reason'; nor is it, 'Can they
talk? but, *Can they suffer?*'[4]

The problems that environmentalists have inherited from the modernist
paradigm become apparent when one turns from the critique of the division
between humanity and nature to the articulation and justification of their vision
of unity. There are three related terms here that correspond to modernist notions
– 'biocentrism', 'ecocentrism' and 'biodiversity'. The first two are principles of
equality. The third often passes for a purely technical term that denotes the
plurality of life forms, the importance of which is explained on mostly
instrumental grounds. Yet, as we have already seen, even in such discourses,
ethical considerations cannot be wholly ignored. I will be using the term here in
its wider cultural sense, which refers to, and celebrates individuality and
difference and as such reproduces the anthropological notion of 'cultural
diversity' at a more inclusive level. 'Respect for biological diversity', says Klaus
Töpfer (UN 1999: xi), Executive Director of UNEP, 'implies respect for human
diversity'. To be sure. Human diversity is an earlier, foundational notion,
extended by environmentalists to cover all forms of life and all kinds of
communities.

'Biocentrism' is an egalitarian attitude. 'However', Fox points out, 'because
the prefix *bio-* refers, etymologically, to life or living organisms, it has sometimes
been assumed that deep ecology's concerns *are* restricted to entities that are …
biologically alive'. Because such is not the case – deep ecology is concerned with
every being and excludes no-thing – Fox and other radical environmentalists
prefer 'to describe the[ir] kind of egalitarian attitude … as *ecocentric* rather than
biocentric'. A broader term to be sure, which is more faithful to the philosophy
of radical ecologies. In addition, it has the advantage of clarifying wider
connections. 'In accordance with this extremely broad, ecocentric egalitarianism,
supporters of deep ecology hold that their concerns well and truly subsume the
concerns of those movements that have restricted their focus to the attainment
of a more egalitarian *human* society … for example … feminism … Marxism,
antiracism, and anti-imperialism'. 'Beyond Humanism', then, but never losing
sight of it, beyond it and further to the other side but only as far as a new
egalitarian attitude – ecocentrism – that both contains Humanism and
reproduces it. Radical environmentalists 'see no *essential* disagreement between
deep ecology and these perspectives, *providing* that the latter are genuinely able
to overcome their anthropocentric legacies' (Fox 1995: 270–71). If there is no
'essential' disagreement, it is only because environmentalism shares with these
paradigms the same 'essence' – a vision of unity, purity and innocence.

There are at least two major problems that environmentalists have inherited
from the modernist paradigm, and these are reflected in the two key

environmentalist notions of 'ecocentrism' and 'biodiversity' – the latter understood in the wider cultural sense as respect for all forms of life. The first is a problem of grounding, the second a problem of contradiction, which manifests itself as a difficulty in reconciling the notions of unity and diversity. As we have seen, many radical environmentalists explain their ecocentric egalitarian attitude on the basis of 'intrinsic value' in nature. The first clause of the deep ecology movement, for instance, reads: 'The well-being and flourishing of human and non-human life on Earth have value in themselves (synonyms: intrinsic value, inherent worth). These values are independent of the usefulness of the non-human world for human purposes' (Naess 1995a: 68). As Zimmerman (1994: 47) notes, however, the argument is highly problematical and burdens deep ecologists with 'the daunting task of proving that nature is intrinsically valuable, that is, valuable apart from the evaluative activity of a conscious being'. A daunting task, no doubt, and a problem that environmentalists have inherited and inadvertently reproduce. Kant came up against it in his ethics but neither he nor any of his followers resolved it satisfactorily. Kant sought to transform subjective valuation into objective reality by universalising the ethical Subject – a move that recalls and repeats the construction of an objective physical reality through the universalisation of the epistemic Subject, the Subject that thinks the world through the pure intuitions of time and space. The Kantian argument, in broad outline, is that 'man' thinks of himself as an end in itself, as having absolute value and worth. Since all 'men' think of themselves in this way, subjective valuation becomes objective reality – the ground that grounds human ethics. As Callicott notes in his discussion of environmentalist ethics, however, Kant's solution to this problem is little more than a tautology: 'Unless I'm missing something, when we look closely at Kant's claim that the "absolute worth" (intrinsic value) of rational beings is objective, all we find is an *assertion* that it is, together with some interesting justificatory clues' (1999: 251). Callicott is not missing anything of substance here. Given the nature of the modernist paradigm – it being an 'internally referential system', as Giddens says – tautology is the order of the day. Modernists posit truths about the world and then attempt to ground them by referring them back to the act of positing them.

Zimmerman and other like-minded environmentalists try to bypass the problem of intrinsic value in nature by abandoning ethics completely. Theirs is an argument that uses culture but only insofar as it leads back to nature – the presumed ultimate ground and guarantor of everything. As we have seen, their claim is that respect for nature should not be the outcome of moral obligation. Once people come to understand the 'true' nature of things – that all is made of the same primordial substance – they would come to identify with everything. Respect and care for non-human beings would be spontaneous and natural. The problem then becomes more manageable – a question of education and enlightenment. From then on, it is a matter of letting nature – in this case, human nature, sentiments such as sympathy and love – to take its own course. And yet if it is a daunting task to prove that value in nature exists independently of the evaluative subject, it is as formidable to demonstrate that reality is how environmentalists of Zimmerman's persuasion describe it, that their ontological description adds nothing to reality, in short, that it reflects passively how reality is in and of itself. This is to say that if value in nature cannot be established as an objective fact neither can, for the very same reasons, the true nature of the world. It is to say that environmentalists of Zimmerman's persuasion have done little more than to repeat the problem on another register. The tautology of establishing the truth by asserting that it exists remains and haunts the environmentalist imagination. The environmentalist ontology is no more securely grounded than the environmentalist ethics.

The second and related problem is one of contradiction and manifests itself in the attempt to reconcile the two key environmentalist notions of unity and diversity. Unity, of course, is the ultimate objective, but how is it to be understood exactly? Certainly, it is not to be understood in the sense of sacrificing the individual, whether human or non-human, for the sake of the collectivity. As we have seen, the perception that this might after all be what environmentalists are striving for has given rise to charges of 'ecofascism'. 'The widening and deepening of the individual selves', says Naess, '*somehow* never makes them into one "mass". Or into an organism in which every cell is programmed so as to let the organism function as one single, integrated being' (1989: 173). It 'somehow' does not make them into one 'mass' because if it did, it would be tantamount to denying these 'selves' their individuality, autonomy and right to be different. Yet despite this individuality, autonomy and difference, it is also the case that all beings embody the *same* 'intrinsic' value and worth or, alternatively, are made of the *same* primordial essence. It is this fundamental sameness that justifies the environmentalist call for respect and care for non-human beings, this very notion that constitutes the rationale of radical and not-so-radical environmentalism. How, then, is one to reconcile these two equally important values? Naess frankly admits that he does not know: 'How to work this out in a fairly precise way I do not know. It is a meagre consolation that I do not find that others have been able to do this in their contemplation of the pair unity-plurality. "In unity diversity!" yes, but how?' (1989: 173). How, indeed?

Let us briefly return to Humanism. As we have seen, nineteenth-century scholars defined it as the capacity to think of oneself as an individual – a unique being. But the capacity itself, they insisted, was not unique; it was universal. Here is, once again, Davies' (1997: 21) formulation of these ideas: 'Humanity is both general and special, common and rare. Each of us lives our human-ness as a uniquely individual experience; but that experience ... is part of a larger, all-embracing humanity, a "human condition"'. Although the modernist draws no such conclusion, it should be apparent that the argument is untenable: it claims that uniqueness is common, individuality collective, difference a form of sameness. Hence, the paradox that plagues Humanist ethics, the contradiction reflected in, among other things, all those tedious, unresolved and un-resolvable debates about the nature of social reality – the debates, that is, which try to determine whether it is the individual or society who must be accorded primacy and which, more often than not, end up proposing an uneasy and inherently unstable compromise. Such is also the contradiction also that environmentalists have inherited from the modernist paradigm and encounter at the level of the individual being, whether animal or plant, vis-à-vis what is in this context society's equivalent – the ecosystem.

In this discussion of homologies, there is, finally, the question of practice to be considered – of radical environmentalist practice and its Humanist precedents. My concern here is not with what is commonly understood as political practice but rather with a more distinctive, and in a certain sense more radical kind. For although no doubt political, the kind of practice that I have in mind is enacted in an idiom that, the modernist would say, undermines political life properly understood. My concern is with religious practices, the ritualised acts with which the 'moral community' celebrates and reproduces itself. Radical environmentalist practices of this sort are still very much in the limelight, their Humanist precedents long buried and largely forgotten. But there are important parallels to be drawn between the two that not only demonstrate similarities in ethos but also suggest that radical environmentalism may be an unmediated expression of a more tentative, less articulate, certainly more constrained, but nonetheless fairly widespread sentiment. Apparently not everyone has reflected systematically on the possible solutions to the current social and environmental problems, much less decided that it is 'man' who is the primary culprit. But it is equally apparent that faith in 'man' is no longer unqualified, support for his projects of mastery less forthcoming. This, of course, is not to say that scepticism and doubt will necessarily crystallise in the form of radical environmentalist theory and practice. It is more likely that radical environmentalism will itself be absorbed, assimilated, institutionalised, normalised and reproduced by wider society in a more domesticated form. To ridicule radical environmentalism, therefore, and to dismiss it as a cultural abomination, as the apologists of the modernist paradigm often do, is to close one's eyes to its possible role as a harbinger of things to come.

As homologies go, ridicule and dismissal were precisely the way in which a new religion was greeted in the nineteenth-century – the 'Religion of Humanity'. Ridicule and dismissal were the *initial* reaction, for as Wright notes about the case of the British press of the time, they soon gave way to respect.[5] The Religion of Humanity was the brainchild of the French thinker and founder of Positivism, Auguste Comte; it was also 'a systematic attempt to found a humanist religion which differed from other forms of religious humanism' (Wright 1986: 3). The major difference was its scientific pretensions. Like all religions, it had a creed, a cult and a code of conduct but these, Comte and his followers claimed, had a scientific basis. The Religion of Humanity was not interested in metaphysical speculations – about the existence of God, for instance, or the possibility of an afterlife. Since such issues could not be empirically verified, they were set aside. The God of this scientific religion was Humanity itself, and its minor deities Humanity's constituent parts – the family, the community, wider society. I shall not venture into a discussion of Comte's philosophical and sociological ideas, since many have come down to us through Durkheim's work and are well known. My concern here, in any case, is with the practical manifestations of the Religion of Humanity, particularly the 'Church of Humanity', which was established in England by Richard Congreve. Nonetheless, one or two comments are necessary.

As I pointed out in the last chapter in relation to Durkheim, if there is anywhere in the world where the social is mistaken for God, this is not 'primitive' societies but the 'secular' West. Given his theory of social evolution, Comte would probably not have agreed that the West 'mis-takes' the social for God. He would have been the first to argue however, that the West should take it as such and treat it accordingly. A comment is also necessary about the question of human unity, since this relates to an issue with which all religions, including the environmentalist 'Religion of Gaia', are concerned, namely, 'the meaning of meaning', as Ferry says. Comte was so obsessed with unity, according to Wright (1986: 20), 'that this desire eventually overcame his principle of verification' – the key principle of Positivism. His main notions in this respect were solidarity with Humanity in the present and continuity with past and future Humanity. 'The man who dares to think of himself independent of others', Comte pointed out, 'cannot even put the *blasphemous* conception into words without immediate self-contradiction, since the very language he uses is not his own'.[6] Following Comte, Durkheim branded the 'blasphemous' kind of individualism 'the profane' and its negation in social unity 'the sacred'. Had he not meant it as an explanation of how 'elementary' societies misconstrue the nature of social reality, he would have been correct. For Comte, unity of Humanity in the present was important because it 'could help mitigate the harshness of the world revealed by science'; unity with past and future Humanity was important too because it could help mitigate the most disturbing aspect of this 'harshness' – death. It provided the believers with an '"ideal resurrection" through the power of the imagination, a "subjective immortality" in which the dead live in the memory of the living' (Wright 1986: 20). Solidarity and continuity, then, were Positivism's answer to the problem of 'the meaning of meaning': solidarity in the face of an indifferent and hostile universe – for that is how science revealed it – and continuity in the face of the ultimate discontinuity. Wright's commentary on the Positivist answer reveals his modernist bias. 'The religion of Humanity', he says, 'offers a comforting fiction in the face of a hostile and meaningless universe' (1986: 21). We have encountered this attitude several times already. It grows out of the untenable conviction that the modernist subjectivity proper – the 'man' who faces the 'fate of the times' like a 'man' – has no need for such comforting metaphysical fictions.

The Religion of Humanity was launched in England by Richard Congreve in January 1859 on the anniversary of Comte's death. The first service was held at his house where he assumed the role of the priest and gave a sermon in which he assured the congregation that they were a Church, since they possessed a faith, even if not yet fully developed liturgy or rituals. In 1867, the London Positivist Society was formed and three years later, the Positivist School opened its doors in Chapel Street in London. Disagreements of various kinds, however, including the dislike of some members for the more overtly ritualised aspects of the gatherings, soon led to schism. One of the first outcomes of this split, says Wright (1986: 82), 'was that the Positivist School became known unequivocally as the Church of Humanity, since Congreve no longer felt inhibited in his role as a priest'. Congreve promised to make the religious aspect of Positivism even more overt through the development of liturgy and ritual for the regular services as well as the administering of Comte's nine sacraments: Presentation (equivalent to baptism), Initiation, Admission (becoming servant of Humanity), Destination (sanctioning of a career choice), Marriage, Maturity, Transformation (naming one's successor) and Incorporation – seven years after one's death in which public opinion judges whether one was worthy to be 'buried in the sacred wood surrounding the Temple of Humanity, symbolising [one's] attainment of subjective immortality' (Wright 1986: 37). The Church of Humanity also developed an ecclesiastical year whose highlight was the Festival of Humanity – with a liturgy of its own and 'graced with an "Advent Collect", looking forward to "the perfect day", when Humanity would be known to all'. Other festivals included the Day of All the Dead, the Festival of Holy Women, the Festival of the Virgin Mother (Humanity) and the Commemoration of Auguste Comte's Birth (Wright 1986: 84). The temple in Chapel Street itself had an altar at the east end, an organ, a lectern and a pulpit and around the walls fourteen busts representing Comte and his thirteen calendar saints: Moses (representing ancient theocracy), Homer (ancient poetry), Aristotle (ancient philosophy), Archimedes (ancient science), Caesar (military civilisation), St. Paul (Catholicism), Charlemagne (Feudal civilisation), Dante (modern epic poetry), Gutenberg (modern industry), Shakespeare (modern drama), Descartes (modern philosophy), Frederic II (modern policy) and Bichat (modern science). The temple was also adorned by a tablet of white marble on which Congreve had engraved in bright green letters various Positivist formulas:

Religion of Humanity
Love for our principle
Order as our basis
Progress for our object
Live for others
(Wright 1986: 79)

Although the services in Chapel Street were never well attended, the Church of Humanity continued to expand by opening branches in other major cities. Notable among these are those in Newcastle and Liverpool, which, as Wright notes, continued to function well into the twentieth century. The church in Newcastle was built of iron, and one of its noteworthy innovations was the addition of a new festival to the Positivist year – the Festival of Machines (Wright 1986: 250). The Church in Liverpool opened a brand new Temple of Humanity in 1913. It is noteworthy because of the elaborate liturgy it had developed. At the Festival of the Virgin-Mother, for instance, the service included an explanation of the difference between the Christian and Humanist Virgins – 'Mary the Virgin-Mother of Jesus was the creation of our Virgin-Mother Humanity'. The service continued with further elaborations on Humanity, which led to 'biblical readings involving Mary, interwoven with the Magnificat sung by women and the Nunc Dimittis sung by men'. There were also quotations from Wordsworth, Sir Walter Scott, Byron, Crashaw and Petrarch. This was followed by 'a sustained meditation on Mary, "who has lived in the heaven of the Human imagination" for centuries, "in company with God the Father, the idealisation of power"', and an exhortation for 'more determined efforts to live with the unseen' – a hardly Positivist attitude – 'with our own saints and angels, with the saints, sages and heroes of all time'. The service ended with 'a prayer, a hymn and an organ voluntary' (Wright 1986: 258–59).

The Religion of Humanity did not withstand the test of time. It was too Victorian, too puritanical in its sexual morality, too intellectual to appeal to the wider public – and it also suffered from the general decline from the 1880s onwards in all forms of church attendance. But perhaps the biggest problem it failed to solve, says Wright (1986: 275), 'was that of death and the demand for personal immortality'. The Religion of Humanity did not survive but Humanity as a religion is very much alive, even if, or rather because it is not recognised as such. The demand for human purity and social innocence was posed again in the twentieth century, only with greater urgency, more consistently, systematically and comprehensively. Commenting on Positivism's wider influence, Wright (1986:1) points out that Comte 'can be shown to have made a significant contribution to the less dogmatic but more pervasive humanism of the twentieth century'. That Humanism has been more pervasive in the twentieth century is surely correct. That it has been less dogmatic than in the nineteenth, surely a matter of interpretation. If dogmatism has anything to do with intolerance,

Wright's argument can itself be shown to be misplaced. This is not because of the normalising Humanistic impositions, such as the culture of 'political correctness'. These and no doubt other impositions are surface manifestations of a prior and more fundamental reality – the proliferation of human divisions in the twentieth century. What this phenomenon demonstrates is that the cultural assumptions which make it possible for the modernist subjectivity to recognise, reveal, discuss, analyse, explain, criticise, in a word, objectify different forms of 'oppression' and turn them into legitimate targets have become even more deeply embedded and firmly entrenched. This proliferation, in short, suggests that the modernist subjectivity has become even more intolerant to anything that does not fit its vision of Pure Humanity.

Humanity as the religion of the modernist subjectivity is currently under threat. Yet the danger is not as grave as the apologists of modernity often claim. What is threatened is merely its autonomy. As we have seen, radical environmentalists argue that their concerns 'well and truly subsume' the concerns of those who strive to eradicate human divisions. Humanity is under threat but only of being incorporated into a grander totalisation and abstraction – Life – life on Earth and the life of the Earth. Radical environmentalists make no pretences about the nature and status of this abstraction. Very much like the disciples of the Religion of Humanity, they are the first to point out that it is a religion in its own right.

Why, asks Bron Taylor (1996: 547), 'should we consider Earth First! a religious movement?'. Because, he answers, 'it manifests all the elements that constitute an emerging religious movement'. Earth First! is the name by which a number of radical environmentalists, operating initially in the United States, identify themselves. It is difficult to say how many. The activists claim that they are not a formal organisation with members, but a movement. More importantly, because they advocate and engage in acts of sabotage – known as 'monkeywrenching' or 'ecotage' – they tend to be secretive about such matters. There are however, groups that call themselves by the same name in England and other European countries, which are quite visible in environmental protests. The website of a journal that goes by the same name provides details about the less secretive aspects of the movement:

> Earth First! was founded in 1979 in response to a lethargic, compromising, and increasingly corporate environmental community. ... We believe in using all the tools in the toolbox, ranging from grassroots organizing and involvement in the legal process to civil disobedience and monkeywrenching. Earth First! is different from other environmental groups. ... First of all, Earth First! is not an organization, but a movement. There are no 'members' of Earth First!, only Earth First!ers. It is a belief in biocentrism, that life, the Earth, comes first, and a practice of putting our beliefs into action.[7]

What, then, are the elements that make Earth First! a religious movement? All religions, Taylor points out, involve myth, symbol and ritual. 'Close observation of Earth First! and of the wider deep ecology movement shows an emerging corpus of myth, symbol and rite that reveals the emergence of a dynamic, new religious movement' (1996: 547). Take, for instance, myth as a story of creation – a cosmogony. The primary cosmogony for Earth First!ers, according to Taylor, is the theory of evolution. Yet evolutionary theory is descriptive and cannot explain why the evolutionary process should be valued. 'This is why so much spirituality gets pulled into the Earth First! movement' (1996: 547). A lot of spirituality 'gets pulled' into the movement precisely because it reveals that life is sacred, and the sacred, of course, is by definition what embodies the highest value possible – and hence what is often asserted dogmatically (cf. Milton 1999).

Or take myth as an eschatology. Earth First!ers have their own 'apocalyptic' eschatology. They believe that industrial society will eventually collapse under its own weight – its ecological sins, one might say – and that out of the rubble a new and very different society will be born.

> Earth First! is radical largely due to [its] apocalyptic worldview: there will be a collapse of industrial society because this society is ecologically unsustainable. After great suffering, if enough of the genetic stock of the planet survives, evolution will resume its natural course. If human beings also survive, they will have the opportunity to re-establish tribal lifeways compatible with the evolutionary future. (Taylor 1996: 552)

Consistent with the belief that society should re-establish 'tribal lifeways', Earth First!ers consider themselves a tribal, 'warrior society' and engage regularly in ritual activities. For instance, 'at meetings held in or near wilderness, they sometimes engage in ritual war or "tribal unity" dances, sometimes howling like wolves'. The participants howl, Taylor goes on to explain, because 'wolves, grizzly bears and other animals function as totems, symbolizing a mystical kinship between the tribe and other creature-peoples'. There are also rituals of 'inclusion' called 'Council of All Beings' whose aim is to connect people spiritually to other beings 'and the entire planet'. During these rituals people get 'possessed by the spirits of non-human entities – animals, rocks, soils and rivers, for example'. These entities can then 'verbalize their hurt at having been so poorly treated by human beings'. At the final stages of the ritual 'the humans seek personal transformation and empowerment, through the gifts of special powers from the non-human entities present in their midst. Ecstatic ritual dance, celebrating inter-species and even inter-planetary oneness, may continue through the night' (Taylor 1996: 549).

Here, then, is another religion – with its own myth, ritual and symbol – a century, give or take, after the Religion of Humanity. What is to become of it, of course, is too early to tell. Yet if the Humanist precedent is anything to go by, the environmentalist religion could well be a harbinger of more things to come – things more normalised and domesticated, less radical and extreme but for this reason also more widespread and entrenched. It could be argued, of course, that the two religions are not comparable, that the environmentalist religion is 'primal' while the Religion of Humanity was 'civil' in character. Yet this would be more an expression of bias than an objection of substance. What is important about these religions is not the particular form through which they articulate themselves but rather what they articulate. My aim in making this juxtaposition, in any case, was partly to show their radicalism – ridiculed and, no doubt, feared in equal measure by the respective pillars of the status quo – and, more importantly, to draw attention to the homologies in guiding assumptions, the concepts used and the visions produced. And this, as another necessary prelude to the argument that has been stated several times already and must now be substantiated. If the environmentalist vision of the world is more objectifying, totalising and unifying than the vision of the modernist paradigm, if the object of environmentalist worship – Life – subsumes (without negating) the modernist object of worship – Humanity – this is only because environmentalists take to the logical and onto-logical conclusion the modernist logic they have inherited – the logic of the Same.

Pure Being

Let us, then, return to the metaphysics of unity and its topography. We are looking for the place from which the last division of the modern – the division between humanity and nature – which is visible on the modernist ground, becomes so blurred that the two merge and emerge as essentially and fundamentally the same. We are looking for the place where looking reveals the world in an extraordinary light, a gentle and diffused light that conceals all the sharp edges and corners of the world and reveals it in its purest form yet – a world of Pure Being, the world of pure beings enacting their innocence. As it will become apparent below, this is also the place that renders this magical vision 'uniquely realistic', as Geertz might say, and therefore compelling to those who have had the revelation.

Before proceeding further, however, it may be pertinent to reiterate one or two things. First, although this vision of unity between humanity and nature is associated with the more radical among environmentalists, it is not restricted to them. Radical environmentalists have done little more than to think systematically, articulate clearly and assert more forcefully the wider claim favoured by politicians and international bureaucrats, namely, that 'we are part of nature and nature is part of us'. It behoves 'mainstream' environmentalism to elaborate on this apparently ontological claim and to justify it. To say, as is often said, that this critical insight about the nature of reality has something to do with the recently acquired ability to see our planet from space is hardly enough. For it is not always clear whether this is meant as statement of fact – that one actually sees humanity in nature and nature in humanity from space – or, as in the case of Lovelock, as a statement about what encouraged the sort of reflection that leads to this conclusion. If it is meant as a statement of fact, it must be one of the most naïve empiricist claims in recent decades and, as we have seen, runs into all sorts of intractable problems. If it is meant to suggest that the view from space acted as a catalyst for reflection, it begs the question. What sort of reflection? If the view from external space encourages reflection, where in internal space must one go to visualise the world as a unity? Indeed, is this space spatial at all?

The second issue to raise here concerns the 'blurring' of the world in environmentalist perception. To return to the spatial metaphor, as practical experience clearly shows, the object of observation becomes blurred either because of distancing or because of proximity. At a certain distance, the items contained in a picture become indistinguishable. But such is also the outcome of proximity, the view from a position very close to it. The 'blurring' of the world and the levelling of its differences through intellectual abstraction and generalisation, which is made possible by the transformation of the modernist subjectivity into the Subject of the world, has been discussed extensively in the last chapter. Here I shall do little more than to show that, ironically but not surprisingly, this is one of the two routes that environmentalists themselves follow to arrive at the point where humanity and nature fuse into a unity of Being. The second case, which depicts the modernist subjectivity as an object in the world among other objects – but can do so, as we shall see, only from the position of the Subject – needs to be demonstrated.

The critical thing to note at this early stage is that proximity to the world, to external and internal nature, is an inextricable part of the modernist logic of the Same. On logical grounds, it is inseparable from distancing. It is a movement along the same axis that must follow the reverse direction from distancing if either of them (and the axis) is to exist. That it has been promoted by the modernist subjectivity as a strategy for arriving, *by way of returning*, to the region of the natural and hence, it is supposed, the real cannot be in doubt either. As we shall see, the Romantic critique of culture as that distancing which alienated Western 'man' from the goodness and innocence of nature is a case in point. And so is anthropology, particularly in its guise as a cultural critique of the West by way of its Others. That proximity has been the primary strategy of Western thought, however, is more debatable. This seems to be Derrida's central claim about Western metaphysics, which he reduces to 'logocentrism' – the priority accorded to speech over writing as the paradigmatic gesture of ensuring the unity of Being. Because Derrida interprets this move as an attempt by the modernist subjectivity to maintain control – which it is, but, as we shall see, only by abandoning control (even if symbolically) on a larger canvas – he loses sight of the reverse and more assertive strategy: how the modernist subjectivity took it upon itself to re-create and guarantee the order and unity of the world by giving priority to 'writing' over 'speech', reason and reflection over emotion and experience, culture over nature. This book has gone some way in providing several examples of this sort. In any case, the point is not to grant priority to one over the other.

What is important rather is to realise that both distancing and proximity are strategic options in the same game, which produce, even if by following different routes, the same result, namely, the modernist subjectivity's 'sure continuance' – its salvation – in the figure of the Same.

Let us turn first to the way in which environmentalists arrive at the Same of humanity and nature – to the extent that this is possible at all – through the distance that the Subject of the world places between itself and the object of perception. As we have seen, beyond their reliance on religious conceptions, environmentalists make extensive use of evolutionary biology and cosmological theory. It may be pertinent, therefore, to examine here the wider assumptions on which these theories are based. We have already encountered the distinction that Collingwood makes between the 'Renaissance' view of nature and the 'Modern' view. The former is the mechanistic perception of nature that environmentalists call 'modern', the latter the perception for which environmentalists have no name, partly because they are implicated in it and partly, one presumes, because they would not wish to be called modernists. Both views, Collingwood argues, are based on analogies. In the first case, nature is understood as a machine, in the second as a domain like human society with its own 'vicissitudes' and therefore history – Natural History. Collingwood goes on to show that this change in analogies has had a number of interrelated and far-reaching consequences, but the most important for the purposes of the present discussion is the dissolution of structure into function. In a machine, structure and function are distinct and the latter presupposes the former – to function, a machine must first be constructed. If nature is not a machine but resembles human society, the distinction is effectively dissolved. There is no structure that it is not also, inextricably, a function. Structure in human society, says Collingwood anticipating the latest in social theory by more than three decades (so-called 'practice theory'), exists only insofar as people behave in certain (relatively fixed and predictable) ways. No behaviour of this sort, no structure. This understanding of nature, Collingwood (1945: 19) goes on to say, necessitates a principle of duration, the 'principle of minimum time'. Since the behaviour of any entity takes place in time, there is a minimum time necessary for that entity to exist. In a shorter period, the entity cannot exist because it has no time to do whatever it is that defines it as such an entity. If there is not enough time for oxygen and hydrogen to combine, there can be no water. This is to say that time is of the essence of the world, not only because things can be known *in* time, as Kant argued, but also because they can be known only in *time*. Time determines not only *what* we perceive but also *how* we perceive it. 'How the world of nature appears to us depends on how long we take to observe it: ... to a person who took a view of it extending over a thousand years it would appear in one way, to a person who took a view of it extending over a thousandth of a second it would appear in a different way' (1945: 23).[8] The first person would see not only a molecule of water but also rain, snow, rivers, lakes and oceans. The second would see none of these things.

The epistemological implication of these assumptions should be clear. To understand the nature of the different beings of the world, one needs time. Even more time – incomparably more – is needed when the aim is to understand the nature of the world itself. The longer one looks, the more one sees how the world behaves, and the more is seen, the better one's understanding of its structure or essence. At the limit, if it were possible to take a view of the world extending to infinity, one would know the absolute truth about reality – the nature of Being itself. And the reverse. An instant – 'not the "instant" of "instantaneous" photography, but a mathematical instant containing no time-lapse at all', says Collingwood (1945: 23) – an instant of this sort, then, would reveal nothing about reality whatsoever, which, as we have seen in the brief discussion on Hegel's thing-in-itself and will elaborate on below, in a strange and paradoxical way, also reveals everything there is to know about Being.

Whether they are aware of it or not, environmentalists operate on the basis of the 'principle of minimum time'. Because they raise the ontological stakes so high, because they are after the truth of reality writ large, they maximise time, stretch it backwards and forwards as far as it can be stretched – indeed, as we shall, beyond its limits. Hence, the reliance on evolutionary biology and cosmological theory, despite their descriptive nature and lack of spirituality. But they also adopt the reverse strategy. Rather than maximise time, they minimise it. They compress it, if not to the point of reducing it to a mathematical instant – for that is not of the human world – to a point where time has no time to enter into the world in the wrong measure and make unrecognisable the truth of the world already available to intuition. Let us follow Spretnak here and call the wrong measure the 'short run', the temporal perspective that, as she argues, is responsible for the fateful misconception that 'man' rather than nature is in control. Environmentalists go back in time to its very beginning as well as forward to its very end because this is the only way to take the very long view that reveals the truth of the world. Alternatively, as we shall see below in detail, they bring the flow of time under strict control because this is the only way to take the kind of very short view that reveals the very same truth in shorthand. What they must *not* do is to be misled into any of the intermediate temporal positions. Here is Spretnak again:

> Newtonian physics can explain the behaviour of matter in a certain *middle range*. At the *subatomic* and *astrophysical* levels, however, Newtonian explanations are inadequate. Similarly, our perceptions at the *gross levels* – that we are all separate from Nature and from each other – are revealed as illusion once we employ the subtle, suprarational reaches of the mind, which can reveal the true nature of being: all is One, all forms of existence are comprised of one continuous dance of matter/energy arising and falling away. (1984: 240; my emphases)

Evolutionary biology and cosmological theory deal, in the first instance, with the beginning – the beginning of Life and the beginning of the Universe respectively. Although both theories are well known, it would be useful to follow environmentalists in these epic journeys to the Origin and explore in detail how the 'true' nature of Being becomes visible.

The Story of the Origin of Life

'In the beginning there was high-energy physics, but during the cooling of the universe we encounter the origins of chemistry' – which is to say, we reach the time when atoms began to combine with one another to form molecules. 'Two kinds of chemistry were needed to get life going: the chemistry that generates the so-called building blocks of life – water, carbon dioxide ... methane, and hydrogen sulfide – and the chemistry that allows these to associate into yet larger assemblies called biochemicals'. The first chemistry occurred at about 4.5 thousand million years ago, the second half a thousand million years later. By chance – and this is a story of nothing but chances and accidents – a 'mutant' biochemical acquired the ability to reproduce itself and to pass the instructions of reproduction on to its products. This was a momentous event of the highest order – for such was the beginning of life. 'Thus life emerged from non-life ... the first progenitor cell from whom all creatures flow' (Goodenough 1998: 18–27).

From a position four thousand million years in the past, then, we see the origin of life – the first progenitor cell. We do not literally see it, of course. As Goodenough notes, we will perhaps never know what happened exactly. Nonetheless, she goes on to say, we know enough about the end result of this process to allow us to work backwards and construct a plausible, if not fully accurate picture – 'a world picture', let us note in passing. But, in fact, we see far more than the progenitor cell. 'We realise that we are connected to all creatures [and] not just in food chains or ecological equilibria'. The connection is not merely one of interdependence; it is also, and more importantly, a connection of substance – of 'deep genetic homology'. 'We share a common ancestor. ... We are connected all the way down'. We are, in short, kin to all creatures. 'I walk through the Missouri woods and the organisms are everywhere. ... I open my senses to them and we connect. I no longer need to anthropomorphise them, to value them because they are beautiful or amusing or important for my survival'. There is no need to humanise them and make them the same because they are already the same. 'I take in the sycamore by the river and I think about its story ... the tiny first progenitor that gave rise to it and me' (Goodenough 1998: 72–74).

The Story of the Origin of the Universe

In the beginning, there was nothing, only a mathematical point. 'We go back 13,500 million years to a time of primordial silence ... of emptiness ... before the beginning of time ... the very ground of all being. ... From this state of immense potential, an unimaginably powerful explosion takes place … energy travelling at the speed of light hurtles in all directions, creating direction, creating the universe'. From a position further back in time, then, some 9.5 thousand million years before the origin of life, we see the Big Bang and the beginning of the Universe. Once again, we do not literally see it but we know enough about the end result of the process to construct a plausible picture – one of the grandest, we might add, of all 'world pictures'. 'All that is now, every galaxy, star and planet, every particle existing comes into being at this great fiery birthing. Every particle which makes up you and me comes into being at this instant and has been circulating through countless forms ever since'. The origin of the universe, then, was in a fundamental sense also the origin of 'you' and 'me'. In form, we may not have been 'you' and 'me' for thousands of million years to come, but in substance, we have been present from the beginning. 'When we look at a candle flame or a star, we see the light of that fireball. [Our] metabolism burns with that very same fire now' (Seed and Fleming 1996: 503). All is One – 'you' and 'me', the sycamore by the river, the river and the pebbles in it, the stars and the galaxies. All is the same because every-thing is a manifestation of the same primordial stuff – 'matter/energy' – the product of a cosmic 'dance', as Spretnak says, in which beings arise and fall away in a never-ending cycle of birth, death and rebirth. 'Countless times in that journey [of evolution] we died to old forms, let go of old ways, allowing new ones to emerge. But nothing is ever lost. Though forms pass, all returns' (Seed and Macy 1996: 502).

Let us note a few things about the Origin. First, its metaphysical status: the Origin of the empirical world, which is nowhere to be found in that world. We have encountered this sort of contradiction several times before and, therefore, I shall not dwell on it for very long. Let us first couch it in terms already used. As we have seen, one of the fundamental insights that the modernist subjectivity has about itself is that it is the product of society and culture. People are born in society and become what they are by virtue of its conditions and conditionings. But if so, how does the modernist subjectivity ever come to acquire this insight? If it too is the product of society, how can it know that it is, in fact, society that makes people? The modernist subjectivity must be unlike other people. It must be able to step outside society, witness the making of people and insert itself back into society to reveal to them the truth of social being. And yet, the modernist subjectivity insists that there is no place outside society, no place that can be known by empirical means and hence no place that it can know. Such a place has long been assigned to the realm of the speculative and the metaphysical. The problem that plagues the story of the Origin is of the same nature. The modernist subjectivity understands itself as the end result of a long evolutionary process that began at some 'point' in the distant past when it itself did not exist. But if it, like everything else, is caught within the evolutionary process, how can it ever come to know of the role of evolution as the creator of all beings? This knowledge can only be the result of an encounter with the eruption of the evolutionary process on the empirical scene, a witnessing of its Origin. But that is impossible. The modernist subjectivity was not 'there'; it did not exist at the 'time' of the Origin. No doubt, it can follow the hints provided by the things of the world and try to trace its steps back to the Origin – work its way backwards, as Goodenough says, and construct a plausible picture of it. But no matter how hard it tries, it will never get 'there'. The Origin forever retreats, says Foucault (1973). The closer one gets to it, the further away it moves because there was no time at the 'time' of the Origin, because since Kant there is nothing outside of time, because the modernist subjectivity, being itself the origin of time, recognises as reality only what exists on the inside of time. If, then, the modernist subjectivity encountered the Origin – and it must have done if it understands itself as the product of evolution – if it did, it is because it ventured beyond itself into the other side of reality – the realm of the metaphysical.

The second thing to note about the Origin is that it is directing us to that which it is not. The Origin must exist, even if it can never be known, because without it nothing else would. But by a related and equally compelling logic so must its other. Without an other, the Origin would be inconceivable. I pointed out in the last chapter that according to one view of God, he creates things directly by simply thinking them. Deified Western 'man' has no such power. He can create things only indirectly, by thinking and naming them in relation to other things. Life emerged from non-life, says Goodenough. No doubt. Being able to identify life presupposes that a decision has been made about the meaning of non-life. Time emerged from a mathematical instant, says cosmological theory. It could not have been otherwise. Without this strange instant, time as we know it, would be meaningless. And so, it becomes possible to raise the question that the story of the Origin does not raise but must assume, namely, the question concerning the vision that the Origin makes possible and *its* other. The vision from the Origin, of course, is a vision of sameness. But if everything is the same at the Origin it must be because of something not of the Origin. There must be an other that resists assimilation, retains its otherness, can never be among the Same, and by virtue of its radical difference makes it possible for the Same to recognise its sameness. What of this other, then? What could it be? How exactly does it render sameness visible to the Same? And where is it to be located? If the Origin is the site that generates sameness, and if this other is radically different from the Same – and it has to be if it possesses the power to make *everything* One – it can only be at the other end of the Origin.

There are hints in the environmentalist literature as to where the other of sameness might lie. As we have seen, 'mainstream' environmentalism claims that having seen our planet from space, we come to realise that we are part of nature and nature is part of us. Although the argument is highly problematical for reasons already discussed, it is important to note that it posits not the Origin but a different site as the condition of possibility of sameness. Here is another hint from Naess's discussion of one of Schumacher's articles. 'Schumacher announces that he will in the article "take an overall view" which "can be obtained only from a considerable height"'. Naess quickly translates Schumacher's statement into his own brand of environmental philosophy – what he calls 'Ecosophy T'. 'In the terminology of Ecosophy T, [a considerable height is] "a considerable depth"' (Naess 1989: 188). Naess, of course, is alluding to the Origin here – the idea that we are connected 'all the way down' – but is the change in terminology and perspective legitimate? It may very well be the case that Schumacher meant exactly what he said – 'considerable height', *not* 'considerable depth'. Let us look at an example that indicates both 'height' and 'depth'. Science, says Callicott, has expanded our worldview. In which direction, backwards or forwards? It would seem in both directions at once.

> The Earth is a 'small planet' in an immense inhospitable universe. We and its other denizens *are*, from a cosmic point of view, *close* kin. And, from the same cosmic point of view, we do in fact depend for our existence – with every breath we take, with every morsel of food we eat – on our fellow voyagers in the odyssey of evolution. (Callicott 1999: 130)

To be sure, the 'odyssey of evolution' directs the reader back to the Origin. But it also points outwards to a location far away from it. 'The Earth is a small planet': if so, it is in the here and now of the statement, not in the there and then of the Origin. To be more precise: it is a small planet in the here and now of the statement as this relates to something far beyond the here and now, and certainly even further beyond the there and then of the Origin. This something, it seems, is not the mathematical point of the origin but the 'immensity' of the universe. Indeed, Callicott is more specific than that. From 'a cosmic point of view', he says, all the denizens of this small planet are close kin. But where is this 'point' from which one can have a view of the entire cosmos? For all the reasons discussed already, it can only be outside of the cosmos, on the other side of time and space. And yet, it might be objected, this is only a figure of speech; it is not meant as a literal statement. No doubt. It can never be a literal statement because there is no such point and no such view in the empirical world. It is a symbolic statement meant to convey the idea of radical difference – so radical that in relation to it all the differences of the 'denizens' of the planet are rendered utterly insignificant and inconsequential or, to put it another way, all the similarities among them are invested with so much significance that they now verge on identity. This is not to say that environmentalists are necessarily aware either of the symbolism of the 'cosmic point of view' or of the relational logic at work here. Often the claim that all is One is taken quite literally. Hence, as we have seen, the struggle to mediate the opposition between diversity and unity. Here is another example of deep-seated ambiguity.

> What identification [with everything that is] should not be taken to mean is *identity* – that I literally *am* that tree over there, for example. What is being emphasized is the tremendously *common* experience that through the process of identification my *sense* of self ... can expand to include the tree even though I and the tree remain physically 'separate' (Fox 1990: 231).

So far, so good – even if the experience may not be so 'tremendously' common. Yet the quotation marks around the word 'separate' are not there by accident. Fox goes on to add, significantly enough, in parentheses: '(even here, however, the word *separate* must not be taken too literally because ecology tells us that my physical self and the tree are physically *interlinked* in all sorts of ways)' (Fox 1990: 231–32). Identity, then, must not be taken literally, but neither should difference.

I shall not speculate on what the other of sameness might be. I shall simply call it the 'End' – and this only as a way of distinguishing it from the Origin. Others might wish to call it different names – death, for instance, an eternal darkness or silence, the unknown or unknowable, non-Being, nothingness. What is significant, in any case, is not the name. Nor is it merely the role that the End plays in making sameness visible to the Same. Equally crucial is the ontological threat that it poses to the modernist subjectivity – crucial because it renders the visible compelling. To know that all beings are in a certain sense the same is not necessarily the profoundest of truths. Everyone with elementary knowledge of biology does. To know this sameness in the face of something that threatens 'the meaning of meaning', however, is to know it with intensity and a sense of urgency that cannot but elevate it to the status of a profound truth. 'Think to your next death. Will your flesh and bones back into the cycle. Surrender. Love the plum worms you will become. Launder your weary being through the fountain of life'. Do so, one is tempted to add, without fear of the End. '[For] nothing is ever lost. Though forms pass, all returns' (Seed and Macy 1996: 502).[9] Such is the promise of sameness. But what if some things *are* forever lost? What is one to do about the nagging suspicion that the End may truly be the end and that there is nothing anyone can do to prevent it from breaking in? In that case, there is at least the comfort of knowing that one is not alone. Here is Goodenough again (1998: 75): 'Blessed be the tie that binds. It anchors us. We are embedded in the great evolutionary story of the planet Earth, the spare, elegant process of mutation and selection and bricolage. And this means that we are anything but alone'. And here is Spretnak:

> In late 1995 astronomers discovered *50 billion more galaxies* than they had known previously. This cosmological event was widely reported in the news media, in the wake of which I heard three paradigmatic responses. The first was from a modern mentality: *So what?* The second from a famous literary scholar of the deconstructionist-postmodern persuasion who felt desponded about our fate as a satellite of a 'banal and decentered star'.... I found this reaction terribly sad and utterly hilarious, so I consulted the most insightful cosmologist I know. I asked Brian Swimme, 'What's the meaning of the discovery of 50 billion "new" galaxies

in the universe?' Without an instant's hesitation, he declared exuberantly, 'We can never be alone again! We have all these relations we didn't know about!' Ah, yes. (1997: 184)

The Origin by itself, then, cannot produce the vision of sameness that environmentalists envision. Although it renders the similarity of all beings inevitable, in and of itself, this is neither here nor there. It did not produce this vision at the time of Darwin, nor for a long time afterwards. Nor did it produce it at the time of the invention of the modernist cosmological theory. Time – the original sin, as we shall see – intrudes and dilutes the similarity of beings into difference, which is also to say that it dilutes the significance of similarity into indifference. What is needed in addition to the Origin is a difference that would make the difference generated by time appear insignificant in comparison, a difference so radical that would make similarity appear as nothing short of identity. What is also needed – and this is why so much spirituality 'gets pulled into' radical environmentalism – is a difference that would raise the ontological stakes of similarity so high as to render the vision of unity 'uniquely realistic', a difference that would imbue unity with so much value as to render it what the highest of all values is – sacred. The difference that does both of these things, by virtue of its radical otherness and the ontological threat it poses to the modernist subjectivity, is the difference of the End.

By spanning the entire horizon of temporality, from the 'time' that there was no time to the 'time' that there is no time, environmentalists encounter both the Origin and the End. They produce the grandest of all 'world pictures', a vision of Pure Being, and secure, to the extent that it can be secured, their 'sure continuance'. Yet this is but one strategy, one route that leads, to the extent that it does, to the Same. To be sure, environmentalists who follow a different route are quick to point out that the figure standing at the end of this epic temporal journey is not the figure of the Same but the persona of 'man'. The argument is not inaccurate, only selectively applied. As we shall see, such is also the persona waiting for anyone returning back into the world, wishing to be nothing more than an object among other objects. I shall examine two paradigmatic examples of the strategy of proximity: ecofeminism and a related argument from anthropology. But first, it is necessary to outline its conditions of possibility. Let us return briefly to Collingwood.

A mathematical instant, as we have seen, is not of this world. It contains no 'time-lapse' at all and hence defeats the principle of 'minimum time' – the bare minimum for anything to emerge in the world. And yet, in a certain fundamental sense, this strange instant contains the truth of the world with which it is pregnant in its purest form. At the mathematical instant of evolutionary biology, that magical time when life emerges from non-life, all life is One. Whatever Life is, it is found in the very narrow space of the first progenitor cell. And so is everything at the mathematical instant of cosmological theory. All that is released by the Big Bang is squeezed so tightly together that it can fit into a 'space' that occupies no space whatsoever. When everything is One, there is nothing to say. Nothing can be said because, as Collingwood points out, there are no particular determinations to be made of any kind – whether quantitative or qualitative, spatial or temporal, material or spiritual. Within the One, nothing can be distinguished from anything else no matter what criterion of distinguishability is used. The One is a completely amorphous, smooth, seamless whole, the point of zero difference, Hegel's Pure Being. Indeed, even this much is already far too much to say about the One. If there are no particular determinations to be made *within* it, there are no particular determinations to make *about* it either. The One cannot be distinguished – to say that it cannot be distinguished from anything else would be redundant, since by definition it is the *only* one. Any way one looks at it, then, the One is another word for Nothing. Hence, the paradox: knowing nothing whatsoever is equivalent to knowing everything there is to know. When everything is One, ignorance is the exactly the same 'thing' as knowledge.

A philosophical mind-game, one might say. Perhaps, but if so, it is a philosophical mind-game with deep cultural roots. It refers, in the first instance, to a mythical time of plenitude and innocence where there was nothing to be known because knowledge had not been invented yet, where the distance, gap, difference between humanity and itself indicated by the breaking up of the self and its consciousness in 'self-consciousness' – which is the condition of possibility of all knowledge – did not exist, where humanity was still One with itself and hence with everything else. And it refers also to all attempts by the modernist subjectivity to return to the vicinity of this mythical time, to minimise the difference between the self of humanity and itself, to make, if not ignorance the same as knowledge, then less knowledge count for more, and more knowledge count for less, to deny the role of the Subject and appear as an object among the Same – the other objects of the world. It refers, in short, to all the modernist strategies of self-effacement and decentring. Let us look briefly at this mythical time and some of its modernist variants.

The Archetype

The place where there was no time, where ignorance was the same as knowledge and a blessing, where humanity was One with itself and hence with everything else, where goodness and innocence reigned even if, or rather because, they did not know themselves as such, where, in short, everything was bathed in the diffused light of absolute purity was the Biblical Garden. Time – and hence difference and hence knowledge – was the original sin. Here are some 'Conjectures' about this pure state of being of humanity and the consequences of falling into time from a rather unlikely source – Kant:

> Before reason arose, there were no commandments or prohibitions, so that violations of these were also impossible. But when reason began to function and, in all its weakness, came into conflict with animality in all its strength, evil necessarily

ensued; and even worse, as reason grew more cultivated, vices emerged which were quite foreign to the state of ignorance and hence of innocence. From the moral point of view, therefore, the first step from this state was a *fall*; and from the physical point of view, this fall was a punishment, for it led to a host of hitherto unknown evils. Thus, the history of *nature* begins with goodness, for it is the *work of God*; but the history of *freedom* begins with evil, for it is the *work of man*. (1970b [1786]: 227)

Kant was no Romantic, of course. The evil of culture is not a timeless substance but an initial condition, a 'punishment'. In time, at the end of the history of 'freedom', 'man' would become deified, and his 'work' – culture – would acquire the goodness of nature. 'Since these abilities [impulses] are adapted to the state of nature, they are undermined by the advance of culture and themselves undermine the latter in turn, until art, when it reaches perfection, once more becomes nature – and this is the ultimate goal of man's moral destiny' (1970b [1786]: 228). Goodness, perfection, purity of Being – all this is timeless and remains stationary, fixed permanently at the same point. It is 'man' that moves, indeed, by this reckoning *must* move. Having fallen into evil, time, difference and an ignorance that can no longer be the same as knowledge even though it tries to pass as such, he must move away and push ahead. He must take it upon himself to finish the journey that he himself began, close the circle and become once again reunited with himself and all other beings. Such is the rationale of the strategy of distancing, of that centring whose ultimate objective is de-centring, of the Subject of the world that, as such a Subject, through its power of reasoning, strives to become once again an object among the Same – the other objects of the world.

Romantic Variant I

And the reverse: the strategy of proximity, of the de-centring of the modernist subjectivity, which, having given up on the idea of redemption through time and history, having denied the role of the Subject, strives to arrive at the Same by way of an urgent return. By this reckoning, the further away 'man' moves from the vicinity of the natural, the goodness of internal and external nature, the worse he becomes. There is no circle here, only the long, monotonous, one-way journey of degeneration. The history of 'freedom' begins with evil and ends in evil, the work of 'man' forever destined to remain a deplorable substitute for the work of God.

The reverse comes at the same time as the strategy of distancing and is bound to it as an alter ego. I shall turn briefly to a paradigmatic exponent, Rousseau, whose work Kant was well aware of.[10] 'Our minds have been corrupted in proportion as the arts and sciences have improved. Will it be said that this is a misfortune peculiar to the present age? No, gentlemen, the evils resulting from our curiosity are as old as the world' (Rousseau 1973 [1750]: 8). The evils resulting from our curiosity go back to the very beginning of the world. For was it not curiosity that caused the fall of 'man'? Was it not the fruit of knowledge that caused his expulsion from the Garden? It goes back to the beginning and comes forward. Take 'man' in historical times, Egypt, or Greece, or Rome, or Constantinople. Was it not philosophy and the fine arts (Egypt), letters and the progress of science (Greece), Ovid, Catullus, Martial 'and the rest of those numerous obscene authors' (Rome) that brought about these nations' downfall? As for Constantinople, 'that metropolis of the Eastern Empire', was it not because it became the refuge of the arts and sciences, which were banished from the rest of Europe 'more perhaps by wisdom than barbarism', that it became also the centre 'of the most profligate debaucheries, the most abandoned villainies, the most atrocious crimes'?

Curiosity, then, the desire to know what 'man' was not meant to know, but also pride and arrogance – such are the vices that caused 'man's' fall 'from that happy state of ignorance, in which the wisdom of providence had placed us'. The difficulty in acquiring knowledge should have served as a warning.

> The thick veil with which [providence] has covered all its operations seems to be a sufficient proof that it never designed us for such fruitless researches. … Let men learn for once that nature would have preserved them from science, as a mother snatches a dangerous weapon from the hands of her child. Let them know that all the secrets she hides are so many evils from which she protects them, and that the very difficulty they find in acquiring knowledge is not the least of her bounty towards them. (Rousseau 1973 [1750: 14)

The 'thick veil' seemed sufficient proof of the futility of research to some. For most, research was the only way forward. Here is a final lament from Rousseau about what has been lost as a result of 'research':

> We cannot reflect on the morality of mankind without contemplating with pleasure the picture of the simplicity which prevailed in the earlier times. This image may be justly compared to a beautiful coast, adorned only by the hands of nature; towards which our eyes are constantly turned, and which we see receding with regret. When men were innocent and virtuous and loved to have the gods for witnesses of their actions, they dwelt together in the same huts; but when they became vicious, they grew tired of such inconvenient onlookers, and banished them to magnificent temples. Finally, they expelled their deities even from these, in order to dwell there themselves; or at least the temples of the gods were no longer more magnificent that the palaces of the citizens. This was the height of degeneracy. (Rousseau 1973 [1750]: 20)

Romantic Variant II (and Anthropological Archetype)

The more one investigates the origins of Humanity, says Herder, the 'closer and closer [one gets] to the *happy clime* where *a single pair of human beings*, under the gentlest influences of *creating Providence* … began spinning the thread'.[11] Should one be embarrassed to talk about these origins? 'The history of the human species' earliest developments, as the oldest book describes it, may sound so *short* and *apocryphal* that we are embarrassed to appear with it before the philosophical spirit of our century which hates nothing more than what is *miraculous* and *hidden* – [but] exactly for that reason it is *true*'. Embarrassment, then, should be set aside.

What were the 'gentle' influences of 'creating Providence' on the first human beings – 'the most natural, the strongest, the simplest!, the eternal foundation for all the centuries of human formation'? Not, to be sure, what the philosophical spirit of Herder's century considered important. They were '*wisdom* instead of science, *piety* instead of wisdom, *love of parents, spouse, and children* instead of politeness and debauchery' (Herder 2002 [1774]: 273-74). Because 'in our philosophical, cold, European world ... we are so incapable of *understanding*!, of *feeling* [these influences] any more ... we *mock, deny* and *misinterpret*'. Take the Orient, for instance. 'We' claim that it is despotic, that 'no *Oriental* as such is yet able hardly *to possess any deep concept of humane, better constitution*'. And yet, is it not the case that before European intervention 'the Oriental with his *sensitive child's sense* [was] the *happiest* and *most obedient student* [of providence]?' And did he not learn better through his 'sensitive sense' than with the miscalculated, 'cold', 'philosophical' European method?

> Have you ever taught a child language from the *philosophical grammar?*, taught him to walk from the most abstract *theory of motion?* Was it necessary, or required, or possible to make the easiest or most difficult duty intelligible to him from a *demonstration* in the *science of ethics?* God precisely be praised! that it was *not required* or *possible*. This sensitive nature, *ignorant* and consequently very curious for everything, *credulous* and hence *susceptible to* any *impression, trustingly obedient* and hence inclined to be led to *everything good*, grasping everything with imagination, amazement, admiration, but precisely in consequence *appropriating* everything *that more firmly* and *wonderfully*. (Herder 2002 [1774]: 278–79)

'The Oriental', then, was the happiest student, learned everything more firmly and 'wonderfully' and did not require European learning. Nor did he require European judgment – 'the *universal, philosophical, human-friendly tone of our century* that grants so gladly to each distant nation, each oldest age, in the world "our own ideal" in *virtue* and *happiness*. Is [there] such a unique judge as to *pass judgement on, condemn*, or beautifully *fictionalise*[12] their ethics according to its own measure alone?' Europeans undertook researches in 'the *progress of the centuries*' and became such a judge. Yet no one should believe in their fictions. They '*exaggerated* or made up facts, understated or suppressed contrary facts, hidden whole sides, taken words for [deeds] ... made up novels ... that no one believes' (2002: 297–98).

Europe invented machines and became a mechanical civilisation. People are cogs in the machine and they grind. 'Alas, they can do nothing but grind, and comfort themselves with *free thinking*. Dear weak, annoying, useless free thinking – substitute for everything that they perhaps needed more: *heart!, warmth!, blood!, humanity!, life!'* (2002: 319). What they need more, however, is also what is increasingly becoming scarce in the rest of the world. 'When has the earth been as universally *enlightened* as now?' wonders Herder rhetorically.

> And how this seems to advance further and further. *Whither* do European colonies not *reach*, and whither *will* they not reach? Everywhere the savages, the more they become fond of our brandy and luxury, become *ripe* for our *conversion* too. Everywhere [they] approach ... *our culture*. Will soon, God help us!, all be human beings *like us*[?]

> *Trade* and *papacy*, how much you have already contributed to this great business! *Spaniards, Jesuits*, and *Dutchmen* – you human-friendly, unselfish, noble, and virtuous nations! – how much has not the *civilisation of humanity* to be grateful to you for already in all parts of the world! (2002: 325)

Phenomenological Variant

We have already encountered Heidegger's critique of what he perceived as the metaphysics of the modernist paradigm – for he, no doubt, excluded his ontology from it – in the discussion on 'The Age of the World Picture'. We have also encountered his vision of a golden age – 'the great age of the Greeks', as he himself says. Here I can do little more than to elaborate further on these key ideas in the context of the strategy of proximity.

'The metaphysics of the modern age begins with and has its essence in the fact that it seeks the unconditionally indubitable, the certain and assured, certainty'; it begins with Descartes's *ego cogito [ergo] sum* (1977a: 82). As we have seen, the desire for certainty produced 'man', the persona that made itself Subject and the world its object, the being that exercises dominion over the world through reason and its handmaidens – science and technology. In treating other beings as mere objects of value, however, which is to say, as things that have value because of, and for him, 'man' has degraded Being itself. Not a new phenomenon, according to Heidegger. This sort of degeneration began a long time ago. 'Already from old, insofar as Being itself has been esteemed at all and thus given worth, it has been despoiled of the dignity of its essence' (1977a: 103). Yet it was not until the arrival of the modern age that this attitude developed fully and became firmly entrenched. During the modern age, even 'the first of beings' has been transformed into a useful thing. 'The ultimate blow against God and against the suprasensory world consists in the fact that God, the first of beings, is degraded to the highest value' (1977a: 105). A degrading of Being, which is also a 'killing'. 'The value-thinking of the metaphysics of the will to power is murderous in the most extreme sense, because it absolutely does not let Being itself take its rise, i.e. come to the vitality of its essence. Thinking in terms of value precludes in advance that Being itself will attain to a coming to presence in its truth' (1977a: 108). The 'killing' of Being by 'man' precludes this possibility 'in advance' because Being is under no obligation to reveal its truth: '[It] harbors itself safely within its truth and conceals itself in such harbouring'. Such is the mystery of Being. This self-concealing is the mystery by means of which 'the truth of Being is coming to presence' (1977a: 110). All the more reason, then, why 'man' should be attentive to Being lest it decides to reveal its truth.

The foreclosing of the possibility of Being ever revealing its truth, which is committed by 'man' through the reduction of all beings to use value, is the essence of European nihilism. Nihilism means that '*Nothing* is befalling Being'. Nothing is befalling it because, in the first instance, 'the truth of Being falls from memory [and] remains forgotten' (1977a: 110). 'Man' has long ceased asking the question concerning the truth of Being, since he thinks he knows what this truth is – value. But it also means, even more fundamentally, that 'Nothing is befalling Being itself. Being *itself* is Being in its *truth*, which truth belongs to Being'. This is to say that nihilism is of the essence of Being. Nothing is befalling it because Being is wrapped up in itself and its truth remains hidden from human eyes, regardless of 'man's' attitude. What are the implications of this peculiarity of Being? First, as we have seen, 'man' should be attentive to Being lest it decides to open itself up and reveal its truth. Second, 'perhaps … we will recognise that neither the political nor the economic nor the sociological, nor the technological and scientific, nor even the metaphysical and the religious perspectives are adequate to think what is happening in [the modern] age'. To borrow Rousseau's expression, all these are 'fruitless researches'. They are because 'what is given to thinking to think is not some deeply hidden underlying meaning, but rather something lying near, that which lies nearest, which, because it is only this, we have therefore constantly already passed over' (1977a: 111).

There are two things to note by way of concluding this brief reference to Heidegger. First, for Heidegger the proximity of Being is demonstrated by the word 'is' – every time people use it to speak about anything, they also speak about Being itself. This is to say, as Heidegger himself says, that 'man' has a pre-reflective understanding of Being. But it is to say also that he understands Being without knowing that he understands it, that his ignorance of Being is also a form of knowledge of what it might be. This is critical in light of the environmentalist strategy of proximity, which I explore below. Second, although Heidegger's discourse is directed against the modernist desire for certainty, it contains important certainties itself: nothing is befalling Being – the truth of Being belongs to Being and to it alone; Being is near by, so near that 'man' constantly passes it over. Because Heidegger uses the term 'nihilism' to name the essence of Being, it is possible to overlook these certainties and to assume that his message is nihilistic. But there is an unmistakable 'making secure' of one's own 'sure continuance' in Heidegger's discourse. What is important is not that the truth of Being belongs to it and it alone. 'Man' may never know this truth, but he can rest assured that it exists and it is very near to him – indeed, it is the 'nearest'.

And so it becomes possible to turn to the environmentalist variants themselves and to raise the question of time once again and, hence, also the question of difference and knowledge. What would be the appropriate amount of time for the unity between humanity and nature to emerge in the world? It cannot be the time of the 'short run' since, as we have seen, this is the source of all error. At this temporal point, the modernist subjectivity has had enough time to investigate and reflect on the world as a totality but not enough time to complete the picture and arrive at the truth. What is missing from the picture is none other than the modernist subjectivity itself. Because it constructs the picture, it forgets to depict itself in it. Hence, the misconception that 'man' is separate from the world. But neither can it be the time of those environmentalists who span the entire horizon of temporality – the 'long run'. Although they construct a picture of unity, the existence of the picture itself belies unity. The issue here is not so much whether these environmentalists remember to insert themselves into the picture as whether getting into it is possible at all. As we have seen, ecofeminists argue that deep ecology's call for identification with all beings falls into the trap of anthropocentrism. Non-human beings are granted moral standing only to the extent that they are incorporated into the expanded human self, which is to make the human being once again, unwittingly no doubt, the centre of the world and the measure of all other beings. There is an argument from environmental anthropology that proceeds along similar lines. 'By anthropocentrism is usually meant an attitude which values all things non-human ... solely as instrumental means. ... Against this, ecocentrism is defined as that attitude which credits the world of nature ... with an intrinsic value. ... Yet despite (or perhaps because of) their conventional opposition, these two positions share more in common than meets the eye' (Ingold 2000: 218). What does not meet the eye, at first sight at any rate, is that ecocentrism is as much of a global perspective as anthropocentrism. It envisions the world as a globe rather than a sphere – the view that, as we have seen in Ingold's discussion of these contrasting perceptions of the world, is responsible for the Western division between humanity and nature. Those environmentalists, then, that construct the world as a 'world picture' remain, necessarily even if unwillingly, outside it forever and in this way reproduce the very dichotomy that they set out to efface.

Not the time of the 'short run', then, and not the time of the 'long run' either. Both are the temporalities of 'man' and hence of 'corruption and degeneration', of culture and hence of the 'heartless, bloodless and lifeless', of theory and hence of the constant 'passing over of that which lies nearest', in short, the temporalities of misunderstanding of the true nature of Being and error. What is needed for the unity between humanity and nature to emerge in the world is far less time. Less but not zero time, not the time of the mathematical instant and its Biblical equivalent, since this is the temporality of ignorance that exists only in pure nature. What is needed is something between these two extremes, not completely of the order of theory and culture but not totally taken over and muted by the silence of pure nature either. What is needed is time enough to generate some difference in the world as well as the knowledge necessary for functioning in it, but not so much as to make either difference or knowledge irredeemable. Time should be such that, as those environmentalists who adopt the strategy of proximity point out, the world would be accessible spontaneously, in an immediate, direct, pre-reflective, pre-objective, pre-ethical, intuitive way; knowledge would be an awareness of the concrete, the particular, the contextual, the experiential – logical enough but as Bourdieu (1980) puts it in another but not unrelated context, logical only to the extent that it does not cease being practical; and although difference would be real, because it would not be burdened by the logic and demands of theory and culture, its boundaries would be porous and permeable. The time in question, is short, would be the time of a world where a quasi-understanding, a fuzzy awareness of the *principle* of unity between humanity and nature would be indicative of the existence of unity and perhaps the best possible guarantee for the continuation of this pure relationship between the two. Let us call this time the 'time of experience' – for such is the limit imposed on it – and let us proceed to explore a few paradigmatic examples.

It should come as no surprise that the time of experience produces a special world, a world of special times, special places, special beings and relationships. Unlike the time of environmentalist theory, which has the time to reflect on the world in its entirety and find the common denominator that makes all beings the same, this time retains all the difference that it inserts into the world. Yet it softens and eventually redeems it. The attitude towards difference in this world is neither one of hostility nor indifference. The otherness of non-human beings is mediated and diffused in proximity, in the personal encounter of the human being with individual non-human beings. It is such encounters that make non-human beings 'special'; and it is through this piecemeal method that a more generalised ethic should be built. Such, in broad outline, is the vision of ecofeminism.

> Special relationship with, care for, or empathy with particular aspects of nature as experiences rather than with nature as an abstraction are essential to provide a depth and type of concern that is not otherwise possible. Care and responsibility for particular animals, trees and rivers that are known well, loved and appropriately connected to the self are an important basis for acquiring a wider, more generalised concern. ... Concern for nature ... should not be viewed as the completion of a process of (masculine) universalization, moral abstraction, and disconnection, discarding the self, emotions, and special ties (all, of course, associated with the private sphere and femininity). (Plumwood 1998: 295)

One begins with experience, then, and on the basis of accumulated experience develops a generalised concern. This is in contrast to the 'masculine' environmentalist method that begins with universalisation and abstraction, discarding in the process the importance of emotions and special ties, and ends up paying the price for it, which is to reproduce all modernist dualisms – first and foremost among them, the division between humanity and nature.

Let us, then, begin with experience ourselves. What is the precise nature of the experience advocated in ecofeminist discourse? To be sure, we know its contents. They are empathy with non-human beings, concern, care, love. This is what people experience in close encounters with particular trees, animals, rivers, rocks; and this is also what they communicate to themselves and others through stories about such encounters. Yet a moment's reflection suggests that ecofeminists, and we through them, know far more than the contents of experience. We know the experience described here as something that speaks about itself – which is not to say because it is narrated but rather because it is narrated and propagated as a means to an end that is not itself. We know it in a wider context than the context of its occurrence or narration, in relation or in opposition to what experience is not – universalisation, moral abstraction, 'disconnection'. This is to say that beyond its contents, we know this experience from the outside, as one possible mode of relating to the world among others. It is to say that the experience propagated by ecofeminism is fully conscious of itself as experience, which makes it something more than itself – a meta-experience. Let us return to the story of the climber and the rock that we encountered in the second chapter.

> I closed my eyes and began to feel the rock with my hands – the cracks and crannies, the raised lichen and mosses, the almost imperceptible nubs that might provide a resting place for my fingers and toes when I began to climb. At that moment I was bathed in serenity. I began to talk to the rock in an almost inaudible, child-like way, as if the rock were my friend. I felt an overwhelming sense of gratitude for what it offered me – a chance to know myself and the rock differently, to appreciate unforeseen miracles like the tiny flowers growing in the even tinier cracks in the rock's surface, and to come to know a sense of *being in relationship* with the natural environment etc. (Warren 1998: 332)

What makes this experience more than itself is not what it says – its contents – nor the fact that is being narrated. Rather, it is what it does not say as a story, what comes before and after it – the pre- and meta-narrative, so to speak. Note that the story is a first-person narrative. 'How is first-person narrative a valuable vehicle of argumentation for ethical decision making and theory building?' asks Warren (1998: 331). As an answer to this question, Warren narrates the climber's story and then goes on to enumerate the advantages of narrating the experience with the rock in the first person. First, Warren says, it 'gives voice to a felt sensitivity; it is a modality which *takes relationships themselves seriously*'. Second, it 'gives expression to a variety of ethical attitudes and behaviors'. Third, 'it provides a way of conceiving of ethics as *emerging out* of particular situations moral agents find themselves in, rather than as being *imposed on* those situations' (Warren 1998: 332–33). Warren goes on to list other advantages but enough has been said to make the point that what we are dealing with here is not merely an experience but one that is fully conscious of itself as an experience – a meta-experience. In and of itself, there is nothing problematical about an experience that has stepped outside itself and is reflecting on what it is and what it can do. The problem is located in the context of the ecofeminist argument, which criticises the 'disconnected, rationalised abstractions' of 'masculinity' on the basis of what is supposed be an engaged, connected, concrete and direct mode of relating to the world – experience. Consider, for instance, the following comments. Plumwood (1998: 303): 'the love of many indigenous people for their land … is based not on vague, bloodless, and abstract cosmological concern but on the formation of identity … in relation to particular areas of land, yielding ties often as special and powerful as those to kin'. Apparently, this is what we should be striving for too – powerful ties that emerge in relation to particular things. And Warren (1998: 341):

> A Sioux elder once told me a story about his son. … The boy was taught, 'to shoot your four-legged brother in his hind area, slowing it down but not killing it. Then take the four-legged's head in you hand, and look into his eyes. The eyes are where all the suffering is. Look into your brother's eyes and feel his pain. Then, take your knife and cut … under his chin … so that he quickly dies. And as you do, ask your brother … for forgiveness … .' As I reflect upon that story, I am struck by the power of the environmental ethic that grows out of and takes seriously narrative [and] context.

Perhaps an environmental ethic does grow out of the two stories about indigenous people. But if it does, it is only for those who have 'grown out' of the experiences related here – the narrative and the context – those for whom narrating the experience is not only a means of describing how they felt but also how they feel about what they felt and what this can do in a society where rationalisation and abstraction are said to have cut many people off from such experiences. The comparison, if that is what it is, between indigenous and ecofeminist experiences is misplaced. For although the contents of the two sets of experiences are similar – both demonstrate the care, concern and respect that people feel about non-human beings in close encounters with them – the experiences themselves are not. As far as we can tell, and if the anthropological literature on the issue is anything to go by, indigenous experiences of non-human beings are known primarily in terms of their content, not as modalities of how people relate to the world, let alone as a possible basis for an environmental ethic.

To articulate its argument, then, ecofeminism must rely on the sort of rationalisation and abstraction that it criticises in 'masculine' discourses, whether environmentalist or otherwise. Its empiricist method of accumulating experiences as the basis for a more generalised concern about nature makes sense and can be effective, indeed, it is possible at all, only against the backdrop of positing in advance what experience as such means and what it can do. With this, we are back to the strategy we have encountered several times already. This is the strategy of the Subject that first posits the meaning of the world – in this case, the meaning of experience as a modality of being in the world – and then pretends, both to itself and to others, to be only an object in the world among other objects. It pretends that its experiences are the same as the experiences of those who do not know of experience as a modality of being. Yet the act betrays itself because eventually it must refer back to, and hence reveal, the positing that makes it possible. This is to say also that the ecofeminist strategy of proximity to internal and hence, external nature as well does not produce the Same but the Same minus the irreducible difference that produces it – the persona of 'man'.

Let us turn to another story about indigenous encounters with non-human beings and what such encounters can teach the West – this time from the most authoritative discourse on the indigenous, namely, anthropology.

> When pursuing reindeer, there often comes a critical point when a particular animal becomes immediately aware of your presence. It then does a strange thing. Instead of running away it stands stock still, turns its head and stares you squarely in the face. … Now the Cree people, native hunters of northeastern Canada … say that the animal offers itself up, quite intentionally and in a spirit of goodwill or even love towards the hunter. The bodily substance of the caribou is not taken, it is *received*. And it is at the moment of the encounter, when the animal stands its ground and looks the hunter in the eye, that the offering is made. (Ingold 2000: 13)

How, asks Ingold, is the Cree claim to be understood? The biologist who already explains the phenomenon as an adaptation to predation by wolfs would probably dismiss it as a fanciful tale. The anthropologist who is not concerned with the truth of the claim but rather its meaning would probably say that in the context of the culture in question, it makes perfect sense – the Cree and other hunters like them believe that non-human beings have intention and agency, that nature is alive. Although the two Western views appear incompatible, Ingold argues, they are in fact complementary. Both are based on a double 'disengagement' of the observer from the world: one that sets up a division between humanity and nature and another that establishes a division within humanity – between enlightened Westerners and ignorant Others. 'First, to suggest [as the anthropologist does] that human beings inhabit worlds of culturally constructed significance is to imply that they have already taken a step out of the world of nature. … The Cree hunter, it is supposed, narrates and interprets his experiences of encounters with animals in terms of a system of cosmological beliefs, the caribou does not'. Narration and interpretation make the Cree hunter a member of humanity, the caribou's apparent inability to do any of the two makes it part of nature. 'But secondly, to perceive this system *as* a cosmology requires that we observers take a further step, this time out of the worlds of culture in which the lives of all other *humans* are said to be confined' (Ingold 2000: 14). The fact that 'we' perceive this system as a cosmology and the fact that we no longer believe in it makes 'us' Westerners members of higher humanity, while the Cree who have this system, believe in what it prescribes but do not know that they have it are relegated to the ranks of lesser humanity.

There is a rather complex chain of interpretations in Ingold's argument. There is first, the Cree hunter's interpretation of his encounter with the caribou, second, the anthropologist's interpretation of the hunter's interpretation and third, Ingold's interpretation of the anthropologist's interpretation of the hunter's interpretation. There is also Ingold's own interpretation of the hunter's interpretation but I shall turn to this below. There is also a set of consequences for the actors involved. The anthropologist's interpretation of the hunter's interpretation places the caribou firmly within nature, the Cree hunter firmly within culture and the anthropologist firmly outside both. But if that is the case, it is necessary to raise the question about the consequences of Ingold's own interpretation of the anthropologist's interpretation. Where does it place him? On the basis of the logic used here, it can only be further afield, at an even greater distance from the world than the distance at which, according to Ingold's argument, the anthropologist places himself. To illustrate this, let us return to Ingold's diagram in his discussion on globes and spheres. If one were to use this diagram to take account of the consequences of Ingold's interpretation of the anthropologist's interpretation, we would have the scheme shown in Figure 3.

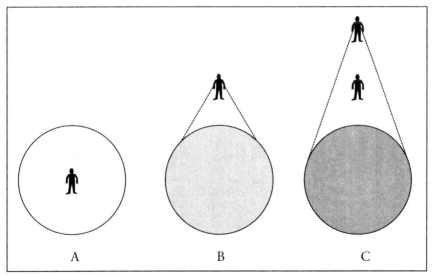

Figure 3 The world as it emerges in: (A) the hunter's interpretation; (B) the anthropologist's interpretation; (C) the anthropologist's interpretation of the anthropologist's interpretation.

I am well aware that the argument developed in this book places me further afield than the anthropologist as well. I am also aware that an interpretation of Ingold's interpretation would place me even further afield. Yet universalisation and abstraction come to bear differently on different discourses. Ingold's aim is to promote an intuitive ecology, a pre-reflexive, pre-objective, pre-ethical mode of relating to the world as a means of doing away with the Western division between humanity and nature. My aim is nothing of the sort.

'I have not forgotten the Cree hunter and the caribou', says Ingold in the concluding section of his paper – a necessary remark, we may note, since he devotes the preceding sections, significantly enough, to a *theoretical* exposition of experience rather than a phenomenological treatment.

> When the hunter speaks of how the caribou presented itself to him, he does not mean to portray the animal as a self-contained, rational agent whose action in giving itself up served to give outward expression to some inner resolution. … The hunter's story is a performance; … its aim is to give form to human feeling – in this case the feeling of the caribou's proximity as another living, sentient being. At that crucial moment of eye-to-eye contact, the hunter *felt* the overwhelming presence of the animal; he felt it as if his own being were somehow bound up or intermingled with that of the animal – a feeling tantamount to love and one that, in the domain of human relations, is experienced in sexual intercourse. In telling of the hunt he gives shape to that feeling in the idioms of speech. (Ingold 2000: 24–25)

One could, of course, raise the question as to how Ingold knows that this is what the Cree hunter *felt*. Yet my intention is not to doubt the authenticity of the experience described here. Nor is it to doubt the authenticity of experience as such. Rather it is to question the claim that the hunter *meant* his story as a performance, that he was narrating it not only for the usual reasons that people have in telling how they felt in particular circumstances but also as an example of how one gives 'form to human feeling'. He may have had such an object in mind, of course – there is no way of knowing. But if so, the hunter's story is more than just a story. It is a story not only about what he felt but also about what a story about feelings means and does – a meta-narrative. As for the experience itself, it is now more than itself. It is a meta-experience, since it is understood against the backdrop of its own definition as experience. It is more likely that the hunter was merely narrating what he felt, not what he meant by what he felt. This, at any rate, is the reason that Ingold narrates the story himself: to demonstrate how people give form to human feeling in a direct, pre-reflective, pre-objective manner, that this is what happens naturally, and that if we think and act in a different manner, it is only because the time of theory and culture – the original sin – is leading us astray.

> When you yell in anger, the yell *is* your anger, it is not a vehicle that *carries* your anger. The sound is not broken off from your mental state and despatched like a message in a bottle cast upon the ocean of sound in the hope that someone might

pick it up. The echoes of the yell are the reverberations of your being as it pours forth into the environment. (Ingold 2000: 24)

It is, of course, true that 'you' do not yell in anger in the hope that someone will pick up the bottle with the message, even though people are known to do this sometimes as well – when they fake anger, for instance. In any case, the important thing is that without anyone to pick up the bottle, whether 'you' or someone else, 'you' would not know why you are yelling. Nor, of course, would 'you' be able to theorise about it, as Ingold does, and argue that 'the yell is your anger', not the meaning attached to it – which is to say that in another and more fundamental sense, the yell is *not* your anger. If it were, if the two were really one and the same thing, 'you' would know neither what 'yell' nor what 'anger' is.

We are, then, once again back to the paradigmatic strategy used by the modernist subjectivity in dealing with itself and the world – the split between the Subject and the object. We are back to the Subject that pretends to be an object among the other objects of the world but betrays itself in the process, the paradoxical situation where to play the role of the object, the modernist subjectivity must first assume the role of the Subject. Hence, the series of contradictions: experience is discussed against the backdrop of its own awareness as experience – 'the yell is your anger'; the critique of disengagement form the world is carried out at the cost of further disengagement – the anthropologist's interpretation of the anthropologist's interpretation places the former at an even greater distance from the world than the latter; a pre-reflective, pre-objective, pre-ethical world is made possible only through theorising, objectifying and moralising about the world. And hence, inevitably, the gap between humanity and nature remains wide open, indeed, now wider than ever before.

During all this time, the apparition of 'man' that hides in the shadows of this irreducible space between humanity and nature has been trying unsuccessfully to suppress its uncontrollable laughter.

Notes

1. Spretnak (1984: 234) quotes the biologist Lewis Thomas in support of her claim: 'Our deepest folly is the notion that we are in charge of the place, that we own it and can somehow run it. We are a living part of Earth's life, owned and operated by the Earth, probably specialized for functions on its behalf that we have not yet glimpsed'.
2. 'Common sense division', that is, in the cultures where environmentalism originates.
3. 'I'm a humanist', says Abbey, one of the founders of the radical group Earth First! 'I'd rather kill a human than a snake' (quoted in *Environmental Ethics*, vol. 4, 1982: 292).
4. Quoted in Singer (1998: 30).
5. The discussion that follows is based on Wright's (1986) study on the 'Religion of Humanity' in Victorian Britain.
6. Quoted in Wright (1986: 24; my emphasis).
7. www.earthfirstjournal.org/efj/primer/index.html

8. Collingwood begins this statement with a caveat: 'The principle [of minimum time] opens no door to subjective idealism'. The water that takes a certain time to form is just as real as the atoms of oxygen and hydrogen of which it is composed.

9. An example of a practical manifestation of this view: 'This sense of identity with other things in nature is partly responsible for a recent trend in burial in the UK. ... Instead of being buried in a conventional coffin, which prevents some agents of decomposition from getting at the body, some people are now being buried in biodegradable coffins which allow their remains to be more quickly absorbed and "recycled" by natural processes, helping to sustain other life forms' (Milton 2002: 155–56; fn. 2).

10. In the 'Conjectures', Kant refers to Rousseau by name – 'the celebrated J.J. Rousseau'. He also refers to Rousseau's 'contradictory pronouncements' (1970b [1786]: 227).

11. All emphases in the quotations from Herder are his.

12. The reference is to classical Greece, '*a discovered favourite people* in antiquity at which we stared until we fell in love with it' (Herder 2002: 295; cf. Bernal 1987 and Herzfeld 1987).

5 No Change

On Hegemony

We are, then, back to the same and the Same. The same is the modernist paradigm and its key figure, the persona of 'man' whose spectre environmentalists believe to have exorcised from their midst but which circulates freely among them and hunts their imagination with its uncanny laughter. It is the logic of the modernist paradigm that reproduces the Same, that system of reasoning which has been trapped by its own cleverness into a monumental naïveté, that way of being which can find its rationale and reason for existence nowhere else except in the Same and hence has never ceased to fantasise about, and strive for unity, human purity and innocence. There is no doubt that at the level of common sense and everyday experience, environmentalism has come a long away from the modernist paradigm of the nineteenth and mid twentieth centuries. But in another and more fundamental sense it has taken not a single step away from it. Environmentalism is the modernist paradigm in its latest guise. It is how modernism now appears having arrived at the limit, how it understands and articulates itself having reached its logical and onto-logical conclusion. With environmentalism, the modernist paradigm has come full circle and has finally consummated itself.

Things could have been otherwise, no doubt, the circumstances and turn of events that have led to the present historical conjuncture different. There is no teleology here, no inevitability, no laws of historical change, whether 'iron' or otherwise. There is only the play of history and chance with a cultural logic, itself a historical product but by now so deeply embedded and entrenched as to constitute a second nature. Environmentalism is not an inescapable outcome of the logic of the Same. But by the same token, it could have arisen nowhere else except in a culture that is driven by this logic.

But there is another reason why environmentalism amounts to a return of the same, and it is with this that I am primarily concerned here. We have encountered this other same early on in the book: how the definition of the meaning of the world was constituted and reproduced in the modernist 'physics' and 'anthropology'; how it was used by European 'man' to legitimise in his own eyes as much as the eyes of the rest of the world the colonial enterprise; how non-European 'man' had come to recognise the European 'man's' vision and himself in it, which is not to say that he only embraced this vision but also contested and rejected in that paradoxical posture of defiance which, as we have seen, can deny and denounce only insofar as it also accepts – an embracing by default and on the rebound; how the modernist 'physics' and 'anthropology' were reinvented and repackaged as modernisation and development in the post war period; how postcolonial 'man' continued to struggle to become a version of European 'man', different, no doubt, but also the same, only to reproduce this difference in the form of an irredeemable Otherness. With environmentalism, we are back to the same vicious circle of identity and power, back to a redefinition of what it means to be and, therefore, a redefinition also of the order of the world and the world order. We are back to that historical conjuncture where non-European 'man' is once again presented with a new reality fait accompli, is asked to take a stance, recognise it, act in accordance with the imperatives that it brings with it – in short, back to the beginning of a new round of 'conversations and conversions', as the Comaroffs would say. Only that this time the 'conversions' have taken place long ago and the 'conversations' can proceed without the complications and misunderstandings of 'first contact'. And just as surely, we are back also to a new round of doubts, questionings, suspicions, negotiations, rejections and denunciations, partial acknowledgements and full recognitions of this new reality – back, that is, to the whole range of attitudes and practices that constitute the complicity by which non-European 'man' contributes to the reproduction of the conditions responsible for what he is striving to change, namely, his inferior position in the 'global hierarchy of value' (Herzfeld 2004).

My aim in this chapter is to explore this latest complicity. I should point out from the outset, however, that this is not a matter of criticising non-European 'man' for being complicit, of assigning blame and allocating responsibility. As we have seen in the discussion on hegemony, the issue is far too complex to be reduced to anything like intention or self-induced ignorance and hence to a question of accountability. There is no 'outside' of hegemony if by hegemony one understands consent based on the socio-historical unconscious – the taken-for-granted, the undisputed and undiscussed, what goes without saying because it appears natural and necessary. There is no outside because by a certain reckoning – one of the most fundamental reckonings of the modernist subjectivity – the socio-historical unconscious is the condition of possibility of all thought and therefore what can never be surpassed by means of reflection. Paradoxically, thinking this unthought is also, at the same time, not thinking it. As we have seen, every time reflection brings to the level of awareness what has eluded it, it is at the expense of generating another unconscious, a new domain of the taken-for-granted, undisputed and undiscussed. This is why resistance to hegemony does little more than to reproduce it further down the road in another guise, why, in the last analysis, it is as good as non-resistance, why those who become aware of their symbolic domination and strive to free themselves from it are no better off than those who are oblivious to it. As the postcolonial experience clearly shows – not to speak of the experience of gender and race inequality – this is why the dominated end up becoming even more entangled and securely trapped.

Such an argument is likely to strike many people as excessively pessimistic, perhaps even reactionary. But if so, it is because such people are more comfortable with a mythology that we have come to know well in this book, because, that is, they cannot stop fantasising about unity, human purity and social innocence – the angelic figure of the Same. And it is as much because they are hopelessly caught up in a trap that we have also come to know very well in this book, the imprisonment in the figure of 'man', that fateful persona that strives to be both the Subject of the world and an object in it and fails miserably in both roles. The socio-historical unconscious, which is responsible for the consent of hegemony, is this very 'man' in the guise of the Subject of the world pronouncing what he is – the product of a society and a history that he does not know because both are anterior to him, but which in a certain sense he also knows. He knows them not in terms of their content to be sure but in terms of their function, their enigmatic ability to make him what he is. For are they not near to him; so near, in fact, that he always already passes them over, as Heidegger says? Are they not what lies nearest? By the same token, the 'thought', the reflection on hegemony that questions this unthought, comes to terms with itself and in doing so withdraws its consent to the dominant, is this 'man' also, this time in the guise of an object in the world. It is this 'man' who, having being the Subject of the world, having defined himself as the product of society and history and hence of himself – for he can no longer imagine himself as being the product of anything else – having, then, made this fateful decision, now, as an object in the world and armed with this prior understanding of himself, strives to discover himself through all those empirical studies that concern themselves with society and history.

It should be apparent that this 'man' will never find his 'true' self, that he would never know exactly who or what he is. His knowledge of himself will always be incomplete, provisional and subject to revisions. Every time he reaches for the unthought of society and history that would reveal to him who he really is, it moves away, closes up on itself or, to use one of Heidegger's (1977a: 110) characteristic expressions, 'harbours itself ... within its truth and conceals itself in such harbouring'. Although nearest to him, although it surrounds him from every possible direction, and although the subject of his empirical investigations and intense reflection, 'man' has *always already* cut himself off from it by making the socio-historical unconscious the condition of possibility of his thought. It should be apparent also that this 'man' will never liberate himself from the forces of society and history that, he now realises, bear down on him so heavily – whether this 'man' is woman, neither man nor woman, the 'man' of another colour, of the dominated social class, of the dominated sections of the dominant class. His freedom will always be incomplete, provisional and unstable, won here lost there, always subject to recall by the powers that be – his fellow 'man' – and as likely to lapse into its contrary by itself. Having decided by and for himself that it is not natural law or divine will that are responsible for his social condition but rather society and history, 'man' must now struggle with them on the basis of his knowledge of society and history. Since he cannot know society and history in their full transparency, however, since his knowledge of them is always provisional and incomplete, every time he strives to deliver himself from them and from his fellow 'man' that dominates him with the use of these forces, he is also at the same time delivering himself over to them. He is always bound to miss something in the process, overlook something, 'forget' some critical issue that would come back to haunt him and play havoc with his efforts to free himself.

This forgetting, it should be emphasised, is not a mere oversight that could be rectified through closer attention or greater reflexivity. It is an irreducible condition. At the most general level, it is the condition of possibility of 'man's' struggle to liberate himself. Without this forgetting, the taking for granted of certain things, he would have no case against the forces of society and history and his fellow 'man', no reason to oppose them, no knowledge that it is they, in fact, which are responsible for his domination. If he does know, it is only because he has already relied on these forces and his fellow 'man' to provide him with the necessary knowledge. If he struggles against them, it is only because he has already taken them into his confidence, turned them into trusted allies, indeed, used them as weapons against themselves. What 'man' forgets, then, and must forget if this charade is to continue, is the paradox in which he is trapped, namely, that he cannot do without the forces of society and history that bear down on him, that he cannot be 'man' without the dominating presence of his fellow 'man', that both are at one and the same time his worse enemies and indispensable friends. This inescapable forgetting is the complicity that unites 'man' with his fellow 'man' in disagreement or, if one prefers, the disagreement that divides him from his fellow 'man' in complicity. Here, too, 'man' has done what he always does, indeed, what he cannot avoid doing – doubled himself, performed both as the Subject of the world and an object in it and failed miserably in both roles. Having made the forces of society and history both the condition of possibility of his domination and the condition of possibility of his liberation, he cannot but struggle for a freedom at once promised and forever deferred.

Such is also the case with colonial and postcolonial 'man'. This is the 'man' who did not know of his status as such a being until his fellow 'man' who came over from Europe revealed the truth to him; who was taught to act like 'man' and has never ceased doing so ever since; who was dragged into this game as much as he willingly joined it himself – a game that, for the reasons that have just been discussed, is always already lost to the 'man' who taught him how to play it. As I have already pointed out, it is not a question of criticising colonial and postcolonial 'man' for having been trapped into this vicious circle or for having trapped himself even more securely through his own devices. We are all trapped in it and we are all contributing, unwittingly and unwillingly, to the circumstances that reproduce this predicament. Nor is it a question of offering remedies that would cure postcolonial 'man' from the syndrome with which he has been afflicted for centuries – his constitution as the negative par excellence, his absence, lack of cultural being, his not being present in the (European) present, in short, his Otherness. There are no such remedies visible on the horizon of the current historical conjuncture. Perhaps there are no remedies at all. Indeed, it could be the case that the very posing of the question of remedies indicates that it is already far too late for anything of the sort to be possible, that it constitutes a forgetting and taken-for-granted which would come to haunt the remedies at the next turn. But although this has not been decided yet and remains and open question, it is certainly not going to be decided by fantasising about the unity of humanity and the unity of humanity and nature nor by being naïvely optimistic, naïvely activistic and naïvely politically correct.

Finally, it is not a question either of simply criticising European 'man' for being what he is. Once again, the complexity of his condition is such that assigning blame in any direct and immediate way would hardly do it justice. As we have seen, if anything, European 'man' does not quite know who he is and hence, in a certain sense, does not quite know what he does either. Nor is he likely to decide with any degree of certainty while being 'man'. His knowledge of himself is bound to be unstable and subject to periodic revisions, indeed, sometimes even complete reversals. His recent and sudden metamorphosis from a subjectivity that defined itself in opposition to nature, divorced and at a distance from it, to a subjectivity that defines itself as part of nature, embedded in it and deeply concerned about its well-being, is a case in point. Yet if criticising European 'man' would, under the circumstances, be rather ingenuous and simplistic, exonerating him because of his quasi-ignorance would be far worse.

My focus in this chapter, at any rate, is not European 'man'. Enough has been said about him already to render him more transparent both in his old and new guise. My aim, rather, is to explore in some detail the complicities of his fellow 'man' in a world increasingly defined and structured according to the 'physics' and 'anthropology' of environmentalism. These are the complicities that at one and the same time allow European 'man' to maintain his position at the centre of the world, as the source of all legitimate signification and hence legitimate power, and help to strengthen further the environmentalist reality currently upon us, even if what this reality means to postcolonial 'man' is not always the same thing as what it means to European 'man'. No doubt, there are points of convergence, chief among them the understanding that the destruction of nature would eventually spell the destruction of humanity as well. But it has to be remembered that environmentalism put an end to the postcolonial nations' fantasy of cruising 'across the centuries' to the European present and drove home the message that if there was to be any movement, this would be of the order of limping rather than leaping. Hence, as we have seen, the initial reaction of the Non-aligned countries was one of deep suspicion and mistrust. Yet environmentalism turned out to be a reality that could not be ignored, avoided or wished away. It had to be confronted and dealt with, its necessity turned into virtue. And it was. As it turned out, and notwithstanding the restrictions it placed on economic development, environmentalism did make a certain sense in the wider political and cultural scheme of things and did make certain things possible within the limits imposed. And so, for reasons that are not quite the same as those of European 'man', postcolonial 'man' speaking for and on behalf of the nation came to embrace environmentalism, sometimes in ways that, ironically, align him more with the more radical environmentalist factions.

The Double Bind

We have already encountered the imperatives of nation building and the way in which they impinged on the postcolonial nations' desire to develop and modernise. We have also encountered the postcolonial nations' reactions to the environmentalist imperatives as they emerged in the early 1970s – reactions guided, on the one hand, by logical conformity to (Western) science and the facts it has produced about the state of the environment and, on the other, logical conformity to (Western) culture and its long-standing truths about what it means to be a human being and a nation in the community of nations. As we have seen, this double conformity involved a double bind, a set of conflicting values manifested in the meaning of poverty and the meaning of environmental pollution. Having being defined as the condition par excellence of the 'under-developed' (Escobar 1995), poverty was for a long time a sign of cultural pollution or, at any rate, indicative of the ignorance of backwardness, while natural pollution was soon to be defined as cultural poverty, indicative of the depravity, greed, arrogance and ignorance of 'man'. From the early 1970s onwards, therefore, postcolonial 'man' was forced as much as he forced himself to tread the fine line between these two sets of assumptions. He continued to push for modernisation and development but he could no longer ignore the environmentalist reality and its wider political and cultural significance. It was important that he redeemed himself from the cultural pollution of poverty but equally important that he did not fall into the cultural poverty of natural pollution. This balancing act, as we have seen, was normalised and formalised in the notion of 'sustainable development'. And this was accomplished with the complicity of the 'developed' countries themselves. They too had a balance to strike. Although it was the 'developed' that had discovered the environment in its complexity and fragility, at the level of national governments and bureaucracies, they remained as mindful as ever of their status as 'developed' and the imperative of retaining their 'pre-eminence' in the community of nations.

'Sustainable development', then, constitutes the latest complicity by which postcolonial 'man', operating at the level of national government, on behalf and for the sake of the nation, contributes to the reproduction of the conditions responsible for his inferior position in the global hierarchy. It reproduces all the assumptions that underscore the notions of modernisation and development and, by the same token, all the assumptions also that make it possible to identify and 'interpellate' him precisely as such a 'man' – the same, no doubt, but not exactly, modernising and developing but not quite in the Western present yet. And it normalises all the assumptions that redefine development in this manner – the assumptions, that is, of the environmentalist paradigm – the new reality with which he is presented *fait accompli* and with which he can do almost everything he wishes except what is perhaps the most critical thing – ignore it. Since he cannot ignore this new reality, 'sustainable development' becomes for postcolonial 'man' the domain in which he must now move, negotiate, manoeuvre, dispute or recognise, in short, try to gain advantage over European 'man'. It structures in advance his various options and renders inevitable the absence of the most fundamental prospect of all, the possibility of winning the game he is playing. The more strategically he operates, the further he disadvantages himself and the more readily he delivers himself over to European 'man'. The better he plays his role as 'sustainable developer', the more rounds he loses in a game that is always already lost.

Let us, then, explore the options and strategies opened to postcolonial 'man', the twists and turns in the struggle for national advantage and prestige in the community of nations, as these emerge in the Statements of Heads of State or Government of the 'developing' countries at the Earth Summit in Rio in 1992. The first thing to note about these statements is the remarkable consistency with which certain themes are raised and treated, as though the representatives of the 'developing' countries or, to use the terminology that statements themselves use, of the 'South', orchestrated their responses in advance. This possibility cannot be completely discounted, of course. Yet even if this is what happened, the use of similar themes is indicative not so much of collusion as of the limits imposed on options and strategies by the same structuring structure – 'sustainable development'. There is, to begin with, a tendency to dwell on the question of environmental destruction and to magnify it. Unlike the statements of the Heads of State or Government of the 'developed' countries, which, although in general acknowledge the existence of environmental problems, emphasise more what can be done about nature rather than what has already been done to it, the statements by the representatives of the South focus on the 'environmental crisis' and do not lose sight of it. They point to environmental problems in their specificity, enumerate and highlight them, linger on or return to them, point an accusing finger to the causes of the problems, call for urgent measures to rectify them. The tendency, in short, is to dramatise the issue and to present it in a way that the more radical among environmentalists would no doubt approve. Let us call this theme, following Lomborg (2001), 'the litany' of environmental dangers.

The second theme that emerges consistently in the statements of the representatives of the South is the question of poverty. The argument here emphasises what was already agreed during the Stockholm Conference 20 years earlier and reiterated subsequently is such documents as the Report of the World Commission on Environment and Development: that the primary cause of environmental destruction in 'developing' countries is poverty itself. There is, finally, a tendency to employ the rhetoric of the more radical environmentalist factions about the ultimate causes of the 'crisis', namely, that these are far deeper than what is usually assumed and that reversal of the 'crisis' requires drastic changes in values and cultural assumptions. Whether the rhetoric is employed in passing or made a more prominent theme, the argument is that the South retains the values and ethos required for the task, since, unlike the North, it has never completely lost touch with its past and the spiritual side of life.

The Litany

'On the threshold of the third millennium, our planet faces exceptionally serious challenges. Aside from poverty they are:

> Air and water pollution;
> Destruction of sites and soils;
> Urban blight;
> Depletion of the ozone layer;
> The greenhouse effect. (UN 1993: 18)

Thus, the President of the Gabonese Republic reiterates the usual environmentalist dangers. A more polemical statement, and a clear reference as to who is primarily responsible for the destruction of the environment, comes from the President of the Council of State and the Council of Ministers of the Republic of Cuba. 'An important biological species is facing the risk of disappearing as a result of the rapid and progressive extinction of its natural living conditions: man'. Having opened his statement with this ominous forecast, Fidel Castro goes on to clarify:

> We have become aware of this problem now, when it is almost too late to prevent it. It is necessary to point out that the consumer societies are mostly responsible for the destruction of the environment. They emerged from the old colonial metropolises and imperial policies, which in turn gave rise to the backwardness and poverty which afflict the overwhelming majority of mankind. ...

> They have poisoned the seas and rivers, they have contaminated the air, they have weakened and perforated the ozone layer, they have saturated the atmosphere with

gases causing variations in climate whose catastrophic consequences are already being felt.

'They' have done all this and much more: 'Forests disappear, deserts spread, thousands of millions of tons of arable land go into the sea every year; numerous species disappear; popular pressure and poverty lead to desperate efforts to survive, even at the expense of nature' (UN 1993: 38). Not only have they done all this, but 'they' have also driven 'us' into poverty and, in 'our' desperate efforts to survive, forced 'us' to do what we would never have dreamt of doing, namely, damage nature. I shall return to the theme of poverty below. Let us here explore a few more statements about the extent of environmental destruction and the primary culprits of this deed.

The Prime Minister of the Kingdom of Nepal: 'Pollution of air and water, industrial and nuclear waste, alarming emission of carbon dioxide, population pressure, deforestation, green-house effect, global warming and soil erosion have gone a long way to distort our climate and threaten to make our Earth unfit for life'. This threat is a recent phenomenon, according to the Prime Minister. 'Fifty years ago, man understood Nature through science and made some miracles of progress. She was stretched but not strained. She could recuperate on her own'. Although 'man' understood nature through science, which produced 'some' miracles of progress – for the Prime Minister presumably not as many as 'man' claims – the overall situation was still redeemable 50 years ago. The real problems emerged when scientific understanding of nature took over completely. 'Today, man has advanced through science to a stage of apparent conquest of Nature. She can no longer recuperate on her own. Conquered, she seems to beckon to us with a vow of terrible vengeance' (UN 1993: 53–54).

The President of the Republic of Kenya: 'The global environmental track since 1972 has not been laudable. … Apart from failing to address fully the original aspirations as charted out in Stockholm, humanity has witnessed the emergence of several environmental "monsters" which literally threaten our existence'. The original Kenyan aspirations expressed in Stockholm have not been fulfilled but nonetheless 'Kenya takes heart in that mankind still has the capacity to mobilise resources to counter environmental problems such as climate change and global warming, ozone layer depletion, mass extinction of biological resources, desertification and the disposal of hazardous wastes and other by-products of development and industrialisation'. (UN 1993: 101)

The President of the Republic of Colombia:

The tremendous environmental problems that the world is now experiencing were not invented by the developing nations. We already know that most of them are caused by the industrialized economies. For decades the rich countries have based their wealth on the unlimited exploitation of natural resources. During this time, they have accumulated an unquantified but undisguisable debt to the planet. (UN 1993: 109)

And the President of the Republic of Indonesia: 'No one can deny that the world is facing increased danger of environmental catastrophe, of diminishing quality of life and a grave threat to the long-term survival of the global ecosystem'. No one can deny the looming catastrophe because such are the facts, the findings of scientists and other experts. 'With the provisions of the Stockholm declaration largely unfulfilled and the dangers to the environment still rapidly escalating, the international community is again urgently reminded of this clear and persistent danger'. It has already been reminded of the danger by the WCED report *Our Common Future*. 'This report, together with the findings of many scientists, scholars and environmentalists, as well as the indispensable work of the United Nations Environment Programme, has underlined the magnitude of the risks and imminent dangers confronting mankind'. The dangers confronting mankind are, of course, by now well known but the President goes on to reiterate some of the most important nonetheless. 'Alarming statistics reflecting wasteful patterns of production and consumption, *inter alia*, resulting in global warming and the progressive depletion of the ozone layer. ... Unless these self-destructive practices are halted or drastically reduced, our planet is doomed to ecological catastrophe. Life as we know it is at stake" (1993: 128).

What emerges, then, from these, and other similar statements is a picture of doom and gloom reminiscent of the radical environmentalists' own apocalyptic vision of the present. In this respect, the contrast with the statements of the representatives of the 'developed' counties could not have been greater. Thus, the Prime Minister of the Portuguese Republic and at the time President of the European Council, speaking on its behalf, in the reference that comes closest to mentioning any environmental problems:

In conclusion, this Summit presents all of us with new opportunities for understanding the dual role of integrating nations ... in the cause of preserving our planet as a pleasant and healthy place for future generations. ... [We] urge all Governments to inspire by their actions the present generation with an ethic of caring for the planet which will allow future generations to inherit the gift of life and to sustain it and pass it on to their successors. (1993: 18)

And the President of the United States of America: 'The Chinese have a proverb: If a man cheats the Earth, the Earth will cheat man. The idea of sustaining the planet so that it may sustain us is as old as life itself'. This as a preamble and a rebuke to those who think that there is something unique about the present historical conjuncture:

Today this old truth must be applied to new threats facing the resources which sustain us all. ... Some find the challenges ahead overwhelming. I believe their pessimism is unfounded.

Twenty years ago, at the Stockholm Conference, a chief concern of our predecessors was the horrible threat of nuclear war – the ultimate pollutant. No more.

> Twenty years ago, some people spoke of the limits to growth. Today, we realize
> that growth is the engine of change and the friend of the environment.
>
> Today, an unprecedented era of peace, freedom and stability makes concerted
> action on the environment possible as never before. ... The United States will work
> to carry forward the promise of Rio. Because as important as the road to Rio has
> been, what matters more is the road from Rio. (UN 1993: 77)

An optimistic message from the President, then, based on several platitudes.
There is nothing wrong with the environment or, at any rate, nothing so wrong
that could not be fixed; the threat of the ultimate pollutant is 'no more' – that is
the critical thing; all other threats can be managed with concerted action; the
United States will do its bit to carry forward this message.

I have already noted that the bleak picture depicted by the representatives of
the South in many ways rivals the radical environmentalists' own apocalyptic
vision of environmental destruction. Yet the representatives of the South are not
radical environmentalists. They are Presidents and Prime Ministers of countries
which until very recently were striving for rapid transition 'across the centuries'
and were convinced that the 'miracles of progress' performed by science and
technology on the other side of time were not 'some', as the Nepalese President
implies in his speech, but very many. Here is the Gabonese President once again:

> In developing countries, particularly in Africa, awareness of our lagging
> development, lack of capital, the desire to develop at all costs and, let it be said,
> strong incentives to do so rapidly, have caused us to acquiesce and even to
> participate in the establishment of a far from rational system of exploiting our
> riches. We are only beginning to grasp the magnitude and consequences of this
> situation. (UN 1993: 19)

The 'developing' countries, then, have really been misled into employing such an
'irrational' system of exploiting their riches. Being under pressure, both by
themselves and outsiders, they 'acquiesced'. Now that the facts about the
environment have become known, they are beginning to see the problem clearly,
in its full extent. And once again, they 'acquiesce'. Acquiescence is, of course, the
hallmark of hegemony. If the foregoing statements are anything to go by, it is also
the essence of the arguments presented by the representatives of the South –
recognition of the reality the new 'reality', which, much like that which preceded
it, is based on facts and appears as the very embodiment of rationality and truth.

We have encountered the argument of facts many times before and although
it cannot be dismissed – for it does account for a certain degree of logical
conformity to the world – it cannot be relied upon as an explanatory tool either.
As we have just seen, the same ecological facts, which for the representatives of the
South constitute an environmental crisis, are understood and treated in a rather
different manner by the representatives of the North. This is to say that the same
way the radical environmentalist position cannot be explained solely on the basis

of ecological facts, neither can the comparable position of the countries of the South. What we are looking for, then, once again is the sense that ecological facts make over and above their pure materiality, the meaning that they pick up once they begin to circulate in social, cultural and political contexts. What is it that has made the notion of an 'environmental crisis' relevant and meaningful to national leaders set steadfastly on the idea of rapid development and eventually brought their arguments in line with those of radical environmentalists who are as steadfastly set on the reverse idea, namely, little or no development – 'zero growth'? How is this unofficial, unrecognised and unlikely alliance to be explained?

There are, of course, several factors that may account for the South's change of attitude if not change of heart. One could point, for instance, to the increasing disillusionment with the prospects for development. As we have already seen, the first Development Decade of the 1960s was a failure and as the Non-aligned pointed out at the Algiers Summit of 1973, the forecasts for the second Decade were extremely pessimistic. As it turned out, the 1980s was as much a 'decade lost for development' as the two earlier ones. Thus, the Prime Minister of Pakistan at the Earth Summit in Rio speaking also on behalf of the Group of 77 ('developing' countries):

> The fundamental cause of the present economic and environmental crisis must be viewed in the context of an unjust world economic order and has contributed to gross imbalances between North and South. If we look back on the 1980s – *the decade lost for development* – we see a grim panorama that encompasses:
>
> Deteriorating terms of trade for commodity and raw material exports;
>
> Heavy debt-servicing and repayment burdens;
>
> Trade barriers and protectionism;
>
> Budgetary austerity;
>
> Wage restraints and monetary discipline. (1993: 153; my emphases)

The list of factors that constitute the 'grim panorama' goes on. Although the North has been encouraging the South to develop, for the Group of 77 it has done little in practical terms to assist in the effort – indeed, as the list itself is meant to suggest, it has done much to prevent it. Hence, the linkage between 'the economic' and 'environmental crisis'. For the Group of 77, there is, indeed, an environmental crisis and it must be dealt with urgently and effectively – for 'life as we know it is at stake'. As much as the South would like to participate in the global effort to save life, however, as much as it would like to contribute to the 'attainment of the noble objectives that have been agreed upon at this summit', it is unable to do so (UN 1993: 154). The 'economic crisis' that plagues the South makes it impossible. 'Is it fair to expect that countries preoccupied with such bleak economic scenarios can give appropriate attention to their problems of the environment?' (UN 1993: 153). It is, of course, not fair. Not that the 'developing' countries do not wish to give their full attention to the environment. On the contrary, they 'take [their] responsibilities seriously. [They] are committed to the implementation of the provisions of Agenda 21'. Indeed, because they are so committed and concerned, they see it as their 'right to demand that [they] be provided with the requisite implements to enable [them] to contribute optimally to preserving the environmental integrity of our planet' (UN 1993: 154–55).

It seems, then, that the fantasy of 'leaping across the centuries' was beginning to be recognised for what it was before the environmental 'crisis' dealt the final blow. At the same time, however, the 'crisis' itself provided new and unexpected opportunities for actually pressing for development in a legitimate if not as grandiose a way. The argument that the 'developing' countries wish to contribute to the global effort to save the planet – indeed, insist that they must – but are unable to do so because of their economic plight arose out of this conjuncture. And so did the related argument that, as we shall see below, poverty contributes in its own way to the deterioration of the environment and must therefore be eradicated. Both rendered the idea of an 'environmental crisis' not only believable as a proposition but also meaningful and relevant. They lent it the kind of gravity that, as we have seen in the discussion on environmentalist concerns about public apathy to ecological facts, leads people to take up an idea and advocate it in earnest.

But there is another way in which the environmental 'crisis' makes sense to the governments of the South, even if this sense is not quite the same thing as the sense it makes to environmentalists. It confirms what the South has always maintained about the North, both about its global hegemony and about its cultural standing. To begin with, and as it should be apparent already from the foregoing statements, the environmental 'crisis' came to be seen as testimony to the gross economic inequality between North and South, the injustice of the world order itself. It was indicative of the affluence, consumerism and wastefulness of the North, which have been made possible at the expense of the South both through the colonial enterprise itself and through the imperialism of later years. Thus, the Prime Minister of Malaysia:

> The poor are not asking for charity. When the rich chopped down their own forests, built their poison-belching factories and scoured the world for cheap resources, the poor said nothing. Indeed they paid for the development of the rich. Now the rich claim a right to regulate the development of the poor countries. And yet any suggestion that the rich compensate the poor adequately is regarded as outrageous. As colonies we were exploited. Now as independent nations we are to be equally exploited. (UN 1993: 233)

Related to the politics of inequality and exploitation is the much older issue of the politics of culture. As we have seen, the imperatives of nation building were such that although the colonised nations recognised European superiority in the 'external domain' of government and industry, they retained for themselves superiority in the 'internal domain' of the 'spirit'. Europe was superior in its material culture, the South in those aspects of culture that may not be quantifiable but are nonetheless as important. As we shall see below in the discussion on the culture of the South, the 'environmental crisis' came to be seen as a confirmation of this nineteenth-century idea. It was both indicative of European material power and the failing of Europe in the spiritual aspects of culture. It highlighted European moral depravity, greed, selfishness, arrogance, and ignorance of deeper and more profound realities than those made available by science – all the shortcomings, that is, that radical environmentalists themselves attribute to 'man'. Hence, the disastrous consequences of the North's use of its material power, the transformation of the 'few miracles of progress' to a haunting nightmare. Over and above any other consideration, then, the 'environmental crisis' makes sense to the South both because it can be used as a weapon against the North in the context of international politics and economics, and because it confirms what the South has always maintained about itself and the North in the context of cultural politics.

Poverty

'If you examine it carefully', said the President of the Republic of Uganda in his own speech at the Earth Summit, 'you will find two groups destroying the environment'.

> The first group is those who are ignorant, who do not know that they are destroying the environment. Then ... there are others who know that they are destroying the environment but who do not have any means to stop destroying the environment; they do not have means to develop alternative sources of energy ... because of necessity.
>
> Then there is a second group of people who destroy the environment. These are the profit-seekers. They seek maximum profits; they do not want to use environmentally clean methods of making money. ...
>
> The first group ... are in the South. The other group ... are mainly in the North. (UN 1993: 23)

Such is the situation. The profit-seekers in the North 'should be able to reduce their greed and make less money'. If they cannot, they should be "discipline[d]. The ignorant everywhere could be 'sensitize[d]'. And those who are not ignorant but destroy the environment out of 'necessity' should be enabled to 'develop alternative technologies' (UN 1993: 23).

The theme of poverty as the way in which the 'developing' countries contribute to the deterioration of the environment despite themselves appears repeatedly in statements of the representatives of the South and is often contrasted to the affluence of the North, which is held to be primarily responsible for the 'environmental crisis'. Thus, the President of the Argentine Republic, but in a more diplomatic vein: 'Both poverty, with its pressing needs, and wealth, with its compulsive habits, lead to unsustainable practices and lifestyles which deplete natural resources to the detriment of the rights of future generations'. On the side of poverty: 'only where certain thresholds of satisfaction of basic needs are reached can efforts be made to conserve natural resources'. In the meantime, 'natural forests are being cleared to make way for subsistence farming and urban populations are swelling'. On the side of affluence: 'in some situations of abundance, the pressures of consumption can have the effect of depleting resources'. Indeed, they can have more consequences than that. 'Intensive chemical use is depleting the ozone layer, especially over our region, and polluting both land and water resources, while excessive energy use is dangerously aggravating the greenhouse effect'. Once again, the lesson to be learnt is clear. 'Once again, the answer is sustainable development' which for the North means curbing its excesses and for the South 'economic and social growth combined with conservation of natural resources' (UN 1993: 30–31).

The President of the United Republic of Tanzania: 'the stark realities of everyday life have shown that environmental degradation in our country is poverty-driven. ... Poverty is both a cause and consequence of environmental degradation' (UN 1993: 186–87). If so, what needs to be done should be obvious. 'In developing countries it is through the process of development that the protection of the "global" environment can be achieved'. By the same token, 'in the developed countries, the challenge is to restructure production and consumption patterns so that they are less wasteful and less environmentally destructive' (UN 1993: 190).

The President of the Republic of Chile: 'For the developing world, environmental protection is closely linked with the fight against poverty.' To drive home his message, the President goes on to make the following comparison: 'We should all realise that phenomena such as air pollution and the greenhouse effect, depletion of the ozone layer, loss of biodiversity, acid rain and toxic waste, and ... phenomena caused not by high levels of consumption but by extreme poverty and underdevelopment, such as hunger, malnutrition and infant mortality ... are equally harmful' (UN 1993: 201).

And the President of the Republic of Zimbabwe: 'I express the readiness of the Government and people of Zimbabwe to assume their responsibilities for the preservation of our planet, our only home.' But before these responsibilities can be assumed, 'the South needs to be economically viable if it is to play its full role in safeguarding the environment. It makes no sense to mount a campaign against deforestation when alternative fuels or resources cannot be made available'. As for the North, it must 'curtail its wasteful production and consumption patterns, which have contributed to the depletion of natural resources and biological diversity in the South ... excessive, life-threatening pollution of the planet's waters, air and protective ozone layer' (UN 1993: 204–205).

The position of the South found some support in the North. Thus, the Prime Minister of the Kingdom of Norway: 'the poor must be brought home from their exile of bondage and humiliation' (UN 1993: 191). And the President of the Republic of Finland: 'Ways must be found to change the tendency for the industrialized countries to keep raising the level of consumption, while the developing countries lag behind or become ever more poor. Poverty is a human tragedy. It is also related to serious environmental problems' (UN 1993: 224).

Thus, it has come to be accepted by virtually everyone that, as the President of Indonesia put it, 'by serving the cause of economic development, we thus serve the cause of the environment' (UN 1993: 130). What in the early 1970s appeared to the Non-aligned as an irreconcilable contradiction, now emerges, with the complicity of both North and South, as a relation of cause and effect. Poverty is not merely cultural pollution, 'bondage and humiliation', but also a cause of natural pollution. Its eradication would benefit all. It would no doubt eliminate 'hunger, malnutrition and infant mortality'; and it would contribute to the global effort to save the planet. But because poverty is above all cultural pollution, its eradication through economic development, even if 'sustainable', would above all benefit the nation and its leaders. It would increase their prestige and consolidate the nation's standing in the community of nations. Here is, once again, the President of Indonesia, this time commenting on the possible perpetuation of the gap between rich and poor: 'Such a course of events would relegate the developing countries to second class status in the community of nations. It does not require much imagination to realise that such a situation could become the seedbed for potential global conflict' (UN 1993: 131). Not much imagination is needed to realise its potential for global conflict because everyone already knows that this game is not so much about 'hunger, malnutrition and infant mortality' *per se* as it is about national power and prestige.

Culture

'We inhabit a single planet but several worlds. There is a world of abundance where plenty brings pollution. There is a world of want where deprivation degrades life'. Thus, the Prime Minister of the Republic of India sums up the current state of affairs in his statement at the Rio Summit. We have encountered both arguments several times already. But there is something else in the Prime Minister's speech that we have not yet encountered and needs to be discussed.

> Countries which are not at a high level of industrial development also have much to offer to this collective effort [to save the planet]. Their people retain a close affinity and kinship with nature and have learnt to make the best use of its resources in areas like traditional and herbal medicine, water-harvesting and management. ... Their life has a large element of contentment, which prevents over-exploitation of resources. What they really need is a decent normal life. (UN 1993: 2)

Countries at a high level of industrial development will lead the effort for developing environmentally-friendly technologies, according to the Indian Prime Minister. Countries at a low level of industrial development, on the other hand, will contribute their ethic of close affinity and kinship with nature, the lifestyle of their people who are largely contented and, therefore, largely impervious to the trappings of consumerism and plenty. They will also make available their traditional knowledge, which, as we have seen, environmentalists have already rationalised, codified and acronymised – TEK (Traditional Ecological Knowledge) or IEK (Indigenous Ecological Knowledge). Thus, the global division of cultural labour, based on the idea that the West is superior in material culture and the East in culture's spiritual aspects, which was first set up in the nineteenth century in the context of an emerging anti-colonial nationalism, is here reproduced more or less intact.

Yet there is a certain inconsistency in the Prime Minister's statement, which reflects the nation's attempt to capitalise on two distinct sources of value as well as the practical difficulties of navigating between the two terms of the notion of 'sustainable development'. On the one hand, the poor are 'largely' content with their lives of close affinity and kinship with nature; one the other, what they 'really' need is a 'decent normal life'. The poor *must* be in need of a 'decent normal life' because the nation must develop and through its economic strength claim, if not 'pre-eminence' as such, at least a 'decent, normal' position in the community of nations. At the same time, the poor *must* be 'largely' content with their lives. This is not only because the idea of simple, contented and spiritually rich lives conforms to the national stereotype of a superior local culture, but also because, since the emergence of the 'environmental crisis', this kind of life has acquired additional cultural value and can now yield even higher symbolic profits. Hence, the Prime Minister, wrapping up his argument:

> Several hundred years ago, poets in India had paid their tributes to the Earth they cherished. They sang:
>
> 'The ocean is your girdle,
>
> Your Bosom the mountains,
>
> Goddess Earth, my obeisance to you,
>
> Forgive me for daring to touch you
>
> With my feet.'
>
> That remarkable reverence for the Earth is what all of us need to imbibe here in Rio. That will impart real meaning to this Earth Summit. (UN 1993: 3)

'We believers', the Crown Prince of the Kingdom of Morocco pointed out in his own statement, 'see the problem of the environment as one of civilization, faith and nature – a divine creature entrusted to humankind ... for safekeeping'.

Although radical environmentalists would not necessarily agree that the role of humankind is that of stewardship, they would certainly have no qualms with the claim that nature is a 'divine creature'. Nor would they disagree with the argument that the problem with the environment is one of 'civilisation' and 'faith'. Their claim too is that there are currently two kinds of civilisation: the utilitarian, materialistic, spiritually impoverished civilisation of the West whose arrogance and greed has led to the destruction of the environment; and the civilisations of the non-West, which retain their ethical, spiritual and religious values intact and have much to teach the former. 'Now more than ever before', the Prince goes on to say, 'the ethical and spiritual dimension must transform the political and economic order'. Now more than ever before because, as we have seen, the argument of the South, as much as that of radical environmentalists, is that what is at stake is nothing less than Life itself. 'That is why an International Seminar on Environmental Ethics and Spirituality was held in Rabat from 28 to 30 April 1992, under the patronage of His Majesty King Hassan II, on man's role and duties towards his natural environment'. The Prince goes on to quote from a royal letter circulated at the Summit, explaining the work of this Seminar:

> The work of the seminar 'places the overall problem of the relationship between humankind and creation in the only context in which an integrated vision is possible – that of the moral responsibility of thinking beings to whom God has given the privilege of inhabiting a marvellous but fragile world and who have a duty to give thanks for it by respecting its sensibilities and by maintaining it as was created by God, beautiful and healthy'. (UN 1993: 75–77)

Once again, radical environmentalists would have no qualm with the gist of the royal message. Once again, the South does not hesitate to use the 'environmental crisis' to capitalise on its presumed spirituality and to claim the moral high ground vis-à-vis the North. 'Permit me', said the President of Pakistan and Chairman of the Group of 77, in another strategic deployment of the spiritualism of the East, to quote two verses from the Holy Qur'an:

> 'Disorder and destruction have appeared on earth and in the oceans due to what the hands of man have done. (Al-Qur'an xxx: 41)'

and

> 'Do good as God has been good to you, and seek not disorder and destruction on earth. (Al-Quran xxiii: 77)' (UN 1993: 152)

Permission was, of course, granted. The East was allowed to score one more point in a game that has always already been lost to the West.

Such, then, are the complicities of the dominated. This is how the environmentalist 'physics' and 'anthropology' are strategically manipulated to serve the interests of the nation, how they are used in the struggle against economic inequality as much as against cultural inequality and national inferiority. Within this context, environmental problems become an 'environmental crisis' of an unprecedented scale, threatening the whole of the planet and Life itself. They do because there is already a crisis in another context, the ongoing struggle on the part of the South to become the North, which is also, inextricably, the struggle of the East to become the West – different in form, no doubt, but the same in cultural value and worth nonetheless. The deeper the 'environmental crisis' appears, the more legitimate the struggle becomes and the greater the chances of success. The more serious the damage caused to nature, the greater the cultural poverty of the West and the deeper the wisdom of the East, which had always known better, even if it had temporarily forgotten, and even if it was 'tricked' into 'acquiescing and even participating' in the struggle to master nature. The greater the urgency with which the 'environmental crisis' must be confronted and dealt with, the greater the urgency also for the South to develop economically, both because this is the only way it would cease contributing to the worsening of the 'crisis' and the only way also it would begin contributing to the 'crisis's' amelioration. This is how postcolonial 'man', thinking, speaking and acting on behalf and for the sake of the nation, reproduces the 'environmental crisis'. No doubt, there are other ways in which postcolonial 'man' sustains the truths of environmentalism – in his capacity, for instance, as a local or global critic and activist, an enlightened intellectual, a member of the cultivated, sophisticated cosmopolitan middle-class – for different reasons and for other stakes than those concerning the status of the nation. Indeed, it may well be the case that some of these reasons are the same as those of European 'man' in his own capacity as a critic and activist, enlightened intellectual, or member of the middle-class. Yet enough has been said here to highlight one aspect at least of the complicities of the dominated and to make the wider point that in this sense too, environmentalism reflects little more than a return of the same.

Is non-European 'man', then, destined to endure another two hundred years of (post)colonial solitude? Perhaps. What this book has tried to show, in any case, is that the solitude on the other side of the divide promises to be as long and far more profound.

Bibliography

Adas, M. 1989. *Machines as the Measure of Men: Science, Technology, and Ideologies of Western Dominance*. Ithaca, NY: Cornell University Press.

Alexander, J.C. 1996. 'Critical Reflections on "Reflexive Modernization"', *Theory, Culture and Society* 13 (4): 133–8.

Anderson, B. 1991. *Imagined Communities*. London: Verso.

Argyrou, V. 1997. '"Keep Cyprus Clean": Littering, Pollution, and Otherness', *Cultural Anthropology* 12 (2): 159–78.

——— 2002. *Anthropology and the Will to Meaning: A Postcolonial Critique*. London: Pluto Press.

——— 2003. '"Reflexive Modernization" and Other Mythical Realities', *Anthropological Theory* 3 (1): 27–41.

Ariès, P. 1983. *The Hour of Our Death*. London: Peregrine Books.

Auerbach, J.A. 1999. *The Great Exhibition of 1851: A Nation on Display*. New Haven, Conn.: Yale University Press.

Beck, U. 1992a. *Risk Society: Towards a New Modernity*. London: Sage.

——— 1992b. 'From Industrial Society to the Risk Society: Questions of Survival, Social Structure and Ecological Enlightenment', *Theory, Culture and Society* 9: 97–123.

Bentham, J. 1907. *Introduction to the Principles of Morals and Legislation*. Oxford: Clarendon Press.

Berlin, I. 1999. *The Roots of Romanticism*. London: Pimplico.

Bernal, M. 1987. *Black Athena: The Afroasiatic Roots of Classical Civilization*. New Brunswick: Rutgers University Press.

Berry, T. 1995. 'The Viable Human', in *Deep Ecology for the 21st Century: Readings on the Philosophy and Practice of the New Environmentalism*, G. Sessions (ed.), pp. 8–18. Boston: Shambhala.

——— 1996. 'Into The Future', in *This Sacred Earth*, R.S. Gottlieb (ed.), pp. 410–14.

Bhabha, H. 1994. *The Location of Culture*. London: Routledge.

Bird-David, N. 1990. 'The Giving Environment: Another Perspective on the Economic System of Gatherer-hunters', *Current Anthropology* 31 (2): 189–96.

Bookchin, M. 1993. 'What is Social Ecology', in *Radical Environmentalism: Philosophy and Tactics*, P. List (ed.), pp. 93–107. Belmont, CA: Wadsworth.

Bourdieu, P. 1977. *An Outline of a Theory of Practice.* Cambridge: Cambridge University Press.

———— 1984. *Distinction: A Social Critique of the Judgement of Taste.* Cambridge, MA: Harvard University Press.

———— 1990. *The Logic of Practice.* Stanford, CA: Stanford University Press.

Brown, M. and May, J. 1989. *The Greenpeace Story.* London: Dorling Kindersley.

Buckle, H.T. 1878. *History of Civilization in England.* Vol. 1. London: Longmans, Green and Co.

Bury, J.B. 1932. *The Idea of Progress: An Inquiry into its Origin and Growth.* New York: Dover Publications.

Callicott, J.B. 1999. *Beyond the Land Ethic: More Essays in Environmental Philosophy.* New York: SUNY Press.

Capra, F. 1996. *The Web of Life: A New Synthesis of Mind and Matter.* London: Flamingo.

Chatterjee, P. 1986. *Nationalist Thought and the Colonial World: A Derivative Discourse.* Minneapolis: The University of Minnesota Press.

———— 1993. *The Nation and its Fragments: Colonial and Postcolonial Histories.* Princeton: Princeton University Press.

Clifford, J. 1986. 'On Ethnographic Allegory', in *Writing Culture: The Poetics and Politics of Ethnography*, J. Clifford and G. Marcus, (eds), pp. 98–121. Berkeley, CA: The University of California Press.

Collingwood, R.G. 1945. *The Idea of Nature.* Oxford: Oxford University Press.

Comaroff, J. and Comaroff, J. 1989. 'The Colonization of Consciousness in South Africa', *Economy and Society* 18: 267–96.

———— 1991. *Of Revelation and Revolution: Christianity, Colonialism, and Consciousness in South Africa,* Volume 1. Chicago: The University of Chicago Press.

Davidson, B. 1978. *Africa in Modern History: The Search for a New Society.* London: Allen Lane.

Davies, T. 1997. *Humanism.* London: Routledge.

Descola, P. 1996. *In the Society of Nature: A Native Ecology in Amazonia.* Cambridge: Cambridge University Press.

Douglas, M. 1966. *Purity and Danger: An Analysis of the Concepts of Pollutions and Taboo.* London: Ark.

———— 1994. *Risk and Blame: Essays in Cultural Theory.* London: Routledge.

Douglas, M. and Wildavsky, A. 1982. *Risk and Culture: An Essay on the Selection of Technological Risks.* Berkeley CA: University of California Press.

Durkheim, E. 1976 [1915]. *The Elementary Forms of the Religious Life.* London: George Allen and Unwin.

———— 1984 [1893]. *The Division of Labor in Society.* New York: The Free Press.

Ehrenfeld, D. 1978. *The Arrogance of Humanism.* New York: Oxford University Press.

Ellen, R.F. 1986. 'What the Black Elk Left Unsaid: on the Illusory Images of Green Primitivism', *Anthropology Today* 2 (6): 8–12.

Escobar, A. 1995. *Encountering Development: The Making and Unmaking of the Third World.* Princeton, NJ: Princeton University Press.

Evans-Pritchard, E.E. 1965. *Theories of Primitive Religion.* Oxford: Clarendon Press.

Ferry, L. 1995. *The New Ecological Order.* Chicago: The University of Chicago Press.

Foucault, M. 1973. *The Order of Things: An Archaeology of the Human Sciences.* New York: Vintage.

Fox, W. 1990. *Toward a Transpersonal Ecology: Developing New Foundations for Environmentalism.* Boston: Shambhala.

———— 1995. 'The Deep Ecology-Ecofeminism Debate and its Parallels', in *Deep Ecology for the 21st Century*, G. Sessions (ed.), pp. 296–89. Boston, MA: Shambala.

Frazer, J. 1950 [1922]. *The Golden Bough.* New York: Collier.

Gandhi, M. 1963 [1910]. *Hind Swaraj (Indian Home Rule)*, in *The Collected Works of Mahatma Gandhi.* Volume 10, pp. 6–68. Delhi: Ministry of Information and Broadcasting.

Geertz, C. 1973. *The Interpretation of Cultures.* New York: Basic Books.

———— 1993. *Local Knowledge.* London: Fontana.

Giddens, A. 1991. *Modernity and Self-identity: Self and Society in the Late Modern Age.* Cambridge: Polity Press.

Goodenough, U. 1998. *The Sacred Depths of Nature.* Oxford: Oxford University Press.

Gore, A. 1992. *Earth in the Balance: Forging a New Common Purpose.* London: Earthscan.

Gottlieb, R.S. (ed.) 1996. *This Sacred Earth: Religion, Nature, Environment.* New York: Routledge.

———— 1996a. 'Introduction: Religion in an Age of Environmental Crisis', in *This Sacred Earth*, pp. 3–14. New York: Routledge.

———— 1996b. 'Spiritual Deep Ecology and the Left: An Attempt at Reconciliation', in *This Sacred Earth*, pp. 516–31. New York: Routledge.

Grove-White, R. 1993. 'Environmentalism: a New Moral Discourse for Technological Society?', in *Environmentalism: The View from Anthropology*, K. Milton (ed.), pp. 18–30. London: Routledge.

Hamilton, R. 1969 [1830]. *The Progress of Society.* London: John Murray.

Harries-Jones, P. 1993. 'Between Science and Shamanism: the Advocacy of Environmentalism in Toronto', in *Environmentalism: The View from Anthropology*, K. Milton (ed.), pp. 43–58. London: Routledge.

Hegel, G.F. 1991 [1894]. *The Philosophy of History*. Buffalo, NY: Prometheus Books.

Heidegger, M. 1977a. *The Question Concerning Technology and Other Essays*. New York: Harper and Row.

——— 1977b. *Basic Writings*. D.F. Krell (ed.). San Francisco: Harper and Collins.

Herder, J.G. 2002 [1774]. *Philosophical Writings*. Cambridge: Cambridge University Press.

Herzfeld, M. 1987. *Anthropology Through the Looking-Glass: Critical Ethnography in the Margins of Europe*. Cambridge: Cambridge University Press.

——— 2004. *The Body Impolitic: Artisans and Artifice in the Global Hierarchy of Value*. Chicago: The University of Chicago Press.

Hodgen, M. 1964. *Early Anthropology in the Sixteenth and Seventeenth Centuries*. Philadelphia: University of Pennsylvania Press.

Hughes, D. 1996. 'From *American Indian Spiritual Ecology*', in *This Sacred Earth*, R.S. Gottlieb (ed.), pp. 131–46. New York: Routledge.

Hume, D. 1977 [1748]. *An Enquiry Concerning Human Understanding*. Indianapolis: Hackett.

Ingold, T. 2000. *The Perception of the Environment: Essays in Livelihood, Dwelling and Skill*. London: Routledge.

Inkeles, A. and Smith, D. 1974. *Becoming Modern: Individual Change in Six Developing Countries*. Cambridge, MA: Harvard University Press.

IUCN UNEP WWF. 1991. *Caring for the Earth: A Strategy for Sustainable Living*. Gland, Switzerland.

James, W. 1961 [1907]. *The Varieties of Religious Experience*. New York: Collier Books.

Jankowitsch, O. and Sauvant, K.P. 1978. *The World Without Superpowers: The Collected Documents of the Non-Aligned Countries*. Vol. 1. Dobbs Ferry, NY: Oceana Publications.

July, R.W. 1968. *The Origins of Modern African Thought*. London: Faber and Faber.

Kant, I. 1934 [1781]. *Critique of Pure Reason*. London: Everyman's Library.

——— 1970a [1784]. An Answer to the Question: 'What is Enlightenment?', in *Kant: Political Writings*, H. Reiss and H.B. Nisbet (eds), pp. 54–60. Cambridge: Cambridge University Press.

——— 1970b [1786]. 'Conjectures on the Beginning of Human History', in *Kant: Political Writings*, pp. 221–34. Cambridge: Cambridge University Press.

LH. Reiss and HB. Nisbet (eds), Leake, J. 2002. 'Eco-heretic beset by Hate Campaign'. *The Sunday Times*. January 13, 1: 4.

Lévy-Bruhl, L. 1926. *How Natives Think*. New York: Knopf.

Lomborg, B. 2001. *The Skeptical Environmentalist: Measuring the Real State of the World*. Cambridge: Cambridge University Press.

Lovelock, J. 2000. *Gaia: A New Look at Life on Earth*. Second edition. Oxford: Oxford University Press.

Mackenzie, J.S. 1907. *Lectures on Humanism*. London: Swan Sonnenschein.

Malinowski, B. 1954 [1925]. *Magic, Science, and Religion*. New York: Doubleday.

Marcus, G. and M. Fischer 1986. *Anthropology as Cultural Critique*. Chicago: The University of Chicago Press.

Merchant, C. 1980. *The Death of Nature: Women, Ecology and the Scientific Revolution*. San Francisco: Harper and Collins.

Milton, K. 1996. *Environmentalism and Cultural Theory: Exploring the Role of Anthropology in Environmental Discourse*. London: Routledge.

———— 1999. 'Nature is Already Sacred', *Environmental Values* 8: 437–49.

———— 2002. *Loving Nature: Towards an Ecology of Emotion*. London: Routledge.

Mol, A. 1996. 'Ecological Modernisation and Institutional Reflexivity: Environmental Reform in the Late Modern Age', *Environmental Politics* 15: 302–23.

Morehouse, W. 1969. *Nehru and Science, the Vision of New India*. New Dehli: Indian Institute of Public Administration.

Morris, B. 1987. *Anthropological Studies of Religion*. Cambridge: Cambridge University Press.

Naess, A. 1989. *Ecology, Community and Lifestyle: Outline of an Ecosophy*. Cambridge: Cambridge University Press.

———— 1993. 'Identification as a Source of Deep Ecological Attitudes', in *Radical Environmentalism: Philosophy and Tactics*, P. List (ed.), pp. 24–92. Belmont, CA: Wadsworth.

———— 1995a. 'The Deep Ecological Movement', in *Deep Ecology for the 21st Century*, G. Sessions (ed.), pp. 64–84. Boston, MA: Shambala.

———— 1995b. 'Self-realization: An Ecological Approach to Being in the World', in *Deep Ecology for the 21st Century*, G. Sessions (ed), pp. 225–39. Boston, MA: Shambala.

Nehru, J. 1961. *The Discovery of India*. Bombay: Asia Publishing House.

North, R. 1995. *Life on a Modern Planet: a Manifesto for Progress*. Manchester: University of Manchester.

Nyerere, J.K. 1973. *Freedom and Development*. Dar Es Salaam: Oxford University Press.

Pagden, A. 1982. *The Fall of Natural Man: The American Indian and the Origins of Comparative Ethnology*. Cambridge: Cambridge University Press.

Passmore, J. 1980. *Man's Responsibilities for Nature: Ecological Problems and Western Traditions*. London: Duckworth.

Plumwood, V. 1998. 'Nature, Self, and Gender: Feminism, Environmental Philosophy, and the Critique of Rationalism', in *Environmental Philosophy*, M. Zimmerman *et al.* (eds), pp. 291–314.

Posey, D.A. 1998. 'The "Balance Sheet" and the "Sacred Balance": Valuing the Knowledge of Indigenous and Traditional Peoples', *Worldviews: Environment, Culture, Religion* 2: 91–106.

Rostow, W. 1960. *The Stages of Economic Growth: a Non-communist Manifesto.* Cambridge: Cambridge University Press.

Rousseau, J.J. 1973 [1750]. The *Social Contract and Discourses.* London: Dent.

Sagan, C. 1980. *Cosmos.* London: Macdonald Futura Publishers.

Sahlins, M. 1985. *How Natives Think: About Captain Cook, for Example.* Chicago: The University of Chicago Press.

Salleh, A. 1984. Deeper than Deep Ecology: The Eco-feminist Connection. *Environmental Ethics* 6: 339–45.

Seed, J. 1996. 'Invocation', in *This Sacred Earth*, R.S. Gottlieb (ed.), pp. 499–500.

Seed, J. and P. Fleming 1996. 'Evolutionary Remembering', in *This Sacred Earth*, R.S. Gottlieb (ed.), pp. 503–506.

Seed, J. and J. Macy 1996. 'Gaia Meditations', in *This Sacred Earth*, R.S. Gottlieb (ed.), pp. 501–502.

Singer, P. 1998. 'All Animals are Equal', in *Environmental Philosophy: From Animal Rights to Radical Ecology*, M. Zimmerman *et al.* (eds), pp. 26–40.

Soper, K. 1986. *Humanism and Anti-humanism.* London: Hutchinson.

Spretnak, C. 1984. 'The Spiritual Dimensions of Green Politics', in *Green Politics*, C. Spretnak and F. Capra, pp. 230–58. London: Paladin Grafton.

——— 1997. *The Resurgence of the Real: Body, Nature, and Place in a Hypermodern World.* New York: Routledge.

Stocking, G. 1987. *Victorian Anthropology.* Boston, MA: Free Press.

Strathern, M. 1980. 'No Nature, no Culture: the Hagen Case', in *Nature, Culture and Gender*, C. MacCormack and M. Strathern (eds), pp. 174–222. Cambridge: Cambridge University Press.

Szerszynski, B. 1996. 'On Knowing What to Do: Environmentalism and the Modern Problematic', in *Risk, Environment and Modernity*, S. Lash, B. Szerszynski and B. Wynne (eds), pp. 104–37. London: Sage.

Taylor, B. 1996. 'Earth First!: From Primal Spirituality to Ecological Resistance', in *This Sacred Earth*, R. S. Gottlien (ed.), pp. 545–57.

Theodossopoulos, D. 2003. *Troubles with Turtles: Cultural Understandings of the Environment in a Greek Island.* Oxford: Berghahn Books.

The Sunday Times 2001. 'Letting the Mob Rule'. January 21, 1: 16.

Trevelyan, C.O. 1838. *On the Education of the Indian People.* London.

Tylor, E.B. 1874. *Primitive Culture.* Vol. 1. New York: Henry Holt.

United Nations. 1951. *Measures for the Economic Development of Under-Developed Countries.* New York.

——— 1963. *Science and Technology for Development: Volume I. World of Opportunity.* New York.

——— 1973. *Report of the United Nations Conference on the Human Environment, Stockholm, 5–16 June 1972.* New York.

———— 1992. *Agenda 21: Programme of Action for Sustainable Development.* New York.

1993. *Report of the United Nations Conference on Environment and Development, Vol. III, Statements made by Heads of State or Government at the Summit Segment of the Conference.* New York.

———— 1999. *Cultural and Spiritual Values of Biodiversity.* Compiled and edited by D.A. Posey. London: Intermediate Technology Publications.

Walsh, B., Karsh, M. and N. Ansell 1996. 'Trees, Forestry, and the Responsiveness of Creation', in *This Sacred Earth*, R.S. Gottlieb (ed.), pp. 423–35. New York: Routledge.

Ward, B. and R. Dubos 1972. *Only One Earth: The Care and Maintenance of a Small Planet.* Harmondsworth: Penguin Books.

Warren, K. 1987. 'Feminism and Ecology: Making Connections', *Environmental Ethics* 9: 3–20.

———— 1998. 'The Power and Promise of Ecological Feminism', in *Environmental Philosophy*, M. Zimmerman *et al.* (eds), pp. 315–44.

Watson, R.A. 1981. 'Misanthropy, Humanity, and the Eco-warriors', *Environmental Ethics* 14: 95.

Weber, M. 1946. *From Max Weber: Essays in Sociology.* Translated, edited and introduced by H.H. Gerth and C. Wright Mills. New York: Oxford University Press.

White, L. 1967. Historical Roots of our Ecologic Crisis. *Science* 155: 1203–7.

Williams, R. 1977. *Marxism and Literature.* Oxford: Oxford University Press.

World Commission on Environment and Development (WCED). 1987. *Our Common Future.* Oxford: Oxford University Press.

Wright, T.R. 1986. *The Religion of Humanity: The Impact of Comtean Positivism in Victorian Britain.* Cambridge: Cambridge University Press.

Zimmerman, M. 1994. *Contesting the Earth's Future: Radical Ecology and Postmodernity.* Berkeley: University of California Press.

Zimmerman, M., Callicott, J.B., Sessions, G., Warren, K. and J. Clark (eds) 1998. *Environmental Philosophy: From Animal Rights to Radical Ecology.* Upper Saddle River, NJ: Prentice Hall.

Index